METHODS IN MOLECULAR BIOLOGY™

Series Editor
John M. Walker
School of Life Sciences
University of Hertfordshire
Hatfield, Hertfordshire, AL10 9AB, UK

For further volumes:
http://www.springer.com/series/7651

G Protein-Coupled Receptor Signaling in Plants

Methods and Protocols

Edited by

Mark P. Running

Department of Biology, University of Louisville, Louisville, KY, USA

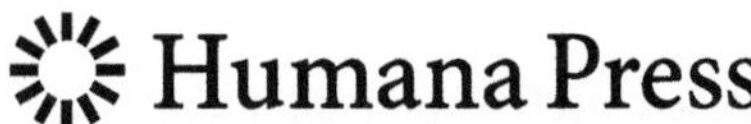
Humana Press

Editor
Mark P. Running
Department of Biology
University of Louisville
Louisville, KY, USA

ISSN 1064-3745 ISSN 1940-6029 (electronic)
ISBN 978-1-62703-531-6 ISBN 978-1-62703-532-3 (eBook)
DOI 10.1007/978-1-62703-532-3
Springer New York Heidelberg Dordrecht London

Library of Congress Control Number: 2013942971

© Springer Science+Business Media, LLC 2013
This work is subject to copyright. All rights are reserved by the Publisher, whether the whole or part of the material is concerned, specifically the rights of translation, reprinting, reuse of illustrations, recitation, broadcasting, reproduction on microfilms or in any other physical way, and transmission or information storage and retrieval, electronic adaptation, computer software, or by similar or dissimilar methodology now known or hereafter developed. Exempted from this legal reservation are brief excerpts in connection with reviews or scholarly analysis or material supplied specifically for the purpose of being entered and executed on a computer system, for exclusive use by the purchaser of the work. Duplication of this publication or parts thereof is permitted only under the provisions of the Copyright Law of the Publisher's location, in its current version, and permission for use must always be obtained from Springer. Permissions for use may be obtained through RightsLink at the Copyright Clearance Center. Violations are liable to prosecution under the respective Copyright Law.
The use of general descriptive names, registered names, trademarks, service marks, etc. in this publication does not imply, even in the absence of a specific statement, that such names are exempt from the relevant protective laws and regulations and therefore free for general use.
While the advice and information in this book are believed to be true and accurate at the date of publication, neither the authors nor the editors nor the publisher can accept any legal responsibility for any errors or omissions that may be made. The publisher makes no warranty, express or implied, with respect to the material contained herein.

Printed on acid-free paper

Humana Press is a brand of Springer
Springer is part of Springer Science+Business Media (www.springer.com)

Preface

Plants, like other multicellular organisms, rely on signal transduction for cell–cell communication and responses to the environment. One prominent mechanism for signaling is the use of small GTPases as on/off switches. In plants, heterotrimeric G proteins, consisting of an alpha, beta, and gamma subunits, and Rop GTPases, which are Rho-related proteins in plants, play a role in a myriad of developmental, hormonal, and environmental responses. In addition, Rab-GTPases regulate intracellular trafficking. This book describes methods used in the study of small GTPases and related proteins.

In eukaryotes, heterotrimeric G proteins are associated with G protein-coupled receptors (GPCRs). Remarkably, while GPCRs are highly prevalent in animals and yeast, very few candidate GPCRs have been identified in plants. One major issue is that GPCRs are not well conserved in sequence. The chapter by Gookin and Bendtsen (Chapter 1) seeks to address this by using bioinformatics approaches to identify proteins matching GPCR characteristics. There is also a comparative dearth in the number of heterotrimeric G protein subunits present in plant genomes. Despite this, heterotrimeric G proteins play roles in several key plant processes, including several described in this book. In contrast, there are several dozen Rab-GTPases present in plants, the function of each of which is just beginning to be understood.

Finally, most G proteins are subject to lipid modification, which serves to facilitate membrane association and to promote protein–protein interactions. Heterotrimeric G protein alpha and gamma subunits, Rop-GTPases, and Rab-GTPases are all subject to the addition to one or more lipid moieties. Several chapters cover how to identify these lipid modifications, which include myristoylation, acylation, and prenylation. Further studies into the roles of small GTPases will help elucidate several key processes in plants.

Louisville, KY, USA *Mark P. Running*

Contents

Contributors

AMIR AKERMAN • *Department of Molecular Biology and Ecology of Plants, George S. Wise Faculty of Life Sciences, Tel Aviv University, Tel Aviv, Israel*
JANNICK D. BENDTSEN • *CLC bio, Aarhus N, Denmark*
JOSE RAMON BOTELLA • *Plant Genetic Engineering Laboratory, School of Agriculture and Food Sciences, University of Queensland, Brisbane, QLD, Australia*
FEDERICA BRANDIZZI • *Department of Energy Plant Research Laboratory, Michigan State University, East Lansing, MI, USA; Department of Plant Biology, Michigan State University, East Lansing, MI, USA*
DAVID CHAKRAVORTY • *Plant Genetic Engineering Laboratory, School of Agriculture and Food Sciences, University of Queensland, Brisbane, QLD, Australia*
JIN-GUI CHEN • *Biosciences Division, Oak Ridge National Laboratory, Oak Ridge, TN, USA*
YANI CHEN • *Department of Energy Plant Research Laboratory, Michigan State University, East Lansing, MI, USA; Department of Plant Biology, Michigan State University, East Lansing, MI, USA*
JEONGMIN CHOI • *Divisions of Plant Science and Biochemistry, University of Missouri, Columbia, MO, USA*
XUEHUI FENG • *Monsanto Company, St. Louis, MO, USA*
TIMOTHY E. GOOKIN • *Department of Biology, Mueller Laboratory, The Pennsylvania State University, University Park, PA, USA*
DIETER HACKENBERG • *Donald Danforth Plant Science Center, St. Louis, MO, USA*
PIERS A. HEMSLEY • *Division of Plant Sciences, College of Life Sciences, University of Dundee, Dundee, UK*
JIRONG HUANG • *National Key Laboratory of Plant Molecular Genetics, Institute of Plant Physiology and Ecology, Shanghai Institutes for Biological Sciences, Chinese Academy of Sciences, Shanghai, China*
INHWAN HWANG • *Division of Integrative Biosciences and Biotechnology, Pohang University of Science and Technology, Pohang, South Korea; Division of Molecular Life Sciences, Pohang University of Science and Technology, Pohang, South Korea*
ZHAOQING JIN • *Biosciences Division, Oak Ridge National Laboratory, Oak Ridge, TN, USA*
LON S. KAUFMAN • *Molecular, Cell, and Developmental Biology, University of Illinois at Chicago, Chicago, IL, USA; Department of Biological Sciences, University of Illinois at Chicago, Chicago, IL, USA*
MYOUNG HUI LEE • *Division of Integrative Biosciences and Biotechnology, Pohang University of Science and Technology, Pohang, South Korea*
YONGJIK LEE • *Division of Integrative Biosciences and Biotechnology, Pohang University of Science and Technology, Pohang, South Korea*
GUOJING LI • *Department of Biology, Hong Kong Baptist University, Kowloon Tong, Kowloon, Hong Kong*

WELLINGTON MUCHERO • *Biosciences Division, Oak Ridge National Laboratory, Oak Ridge, TN, USA*
DANIELLE A. OROZCO-NUNNELLY • *Molecular, Cell, and Developmental Biology, University of Illinois at Chicago, Chicago, IL, USA; Department of Biological Sciences, University of Illinois at Chicago, Chicago, IL, USA*
SONA PANDEY • *Donald Danforth Plant Science Center, St. Louis, MO, USA*
XINGYUN QI • *Department of Biology, McGill University, Montreal, QC, Canada*
SWARUP ROY CHOUDHURY • *Donald Danforth Plant Science Center, St. Louis, MO, USA*
MARK P. RUNNING • *Department of Biology, University of Louisville, Louisville, KY, USA*
WAN SHI • *Washington University in Saint Louis, St. Louis, MO, USA*
NADAV SOREK • *Energy Biosciences Institute, Berkeley, CA, USA; Department of Plant and Microbial Biology, University of California, Berkeley, CA, USA*
GARY STACEY • *Divisions of Plant Science and Biochemistry, University of Missouri, Columbia, MO, USA*
KIWAMU TANAKA • *Divisions of Plant Science and Biochemistry, University of Missouri, Columbia, MO, USA*
YURI TRUSOV • *Plant Genetic Engineering Laboratory, School of Agriculture and Food Sciences, University of Queensland, Brisbane, QLD, Australia*
DONGLI WAN • *Department of Biology, Hong Kong Baptist University, Kowloon Tong, Kowloon, Hong Kong*
XUEJUN WANG • *Monsanto Company, St. Louis, MO, USA*
XUEMIN WANG • *National Key Laboratory of Crop Genetic Improvement, Huazhong Agricultural University, Wuhan, China; Donald Danforth Plant Science Center, St. Louis, MO, USA*
KATHERINE M. WARPEHA • *Molecular, Cell, and Developmental Biology, University of Illinois at Chicago, Chicago, IL, USA; Department of Biological Sciences, University of Illinois at Chicago, Chicago, IL, USA*
COREY S. WESTFALL • *Department of Biology, Washington University, St. Louis, MO, USA*
WENJUAN WU • *National Key Laboratory of Plant Molecular Genetics, Institute of Plant Physiology and Ecology, Shanghai Institutes for Biological Sciences, Chinese Academy of Sciences, Shanghai, China*
YIJI XIA • *Department of Biology, Hong Kong Baptist University, Kowloon Tong, Kowloon, Hong Kong*
SHAUL YALOVSKY • *Department of Molecular Biology and Ecology of Plants, George S. Wise Faculty of Life Sciences, Tel Aviv University, Tel Aviv, Israel*
QIN ZENG • *Monsanto Company, St. Louis, MO, USA*
SHUQUN ZHANG • *Department of Biology, Hong Kong Baptist University, Kowloon Tong, Kowloon, Hong Kong*
XIUJUAN ZHANG • *Department of Biology, Hong Kong Baptist University, Kowloon Tong, Kowloon, Hong Kong*
JIAN ZHAO • *National Key Laboratory of Crop Genetic Improvement, Huazhong Agricultural University, Wuhan, China*
HUANQUAN ZHENG • *Department of Biology, McGill University, Montreal, QC, Canada*

Chapter 1

Topology Assessment, G Protein-Coupled Receptor (GPCR) Prediction, and In Vivo Interaction Assays to Identify Plant Candidate GPCRs

Timothy E. Gookin and Jannick D. Bendtsen

Abstract

Genomic sequencing has provided a vast resource for identifying interesting genes, but often an exact "gene-of-interest" is unknown and is only described as putatively present in a genome by an observed phenotype, or by the known presence of a conserved signaling cascade, such as that facilitated by the heterotrimeric G-protein. The low sequence similarity of G protein-coupled receptors (GPCRs) and the absence of a known ligand with an associated high-throughput screening system in plants hampers their identification by simple BLAST queries or brute force experimental assays. Combinatorial bioinformatic analysis is useful in that it can reduce a large pool of possible candidates to a number manageable by medium or even low-throughput methods.

Here we describe a method for the bioinformatic identification of candidate GPCRs from whole proteomes and their subsequent in vivo analysis for G-protein coupling using a membrane based yeast two-hybrid variant (Gookin et al., Genome Biol 9:R120, 2008). Rather than present the bioinformatic process in a format requiring scripts or computer programming knowledge, we describe procedures here in a simple, biologist-friendly outline that only utilizes the basic syntax of regular expressions.

Key words G-protein coupled receptor, GPCR, G-protein, Bioinformatic, Membrane proteins, Topology, Transmembrane domains, Signal peptide, Yeast two-hybrid, Split-ubiquitin

1 Introduction

The ability to sense and respond to changing environmental conditions is critical for organismal success. In metazoa, the heterotrimeric G-protein complex signaling cascade plays major physiological roles and consists of a membrane localized G-protein complex coupled receptor (GPCR) containing seven transmembrane domains (TMs), a membrane-proximal heterotrimeric G-protein complex consisting of a single Gα, Gβ, and Gγ subunit, and a host of downstream effectors [1]. This evolutionarily conserved signaling mechanism is well conserved in plants [2, 3], yet in comparison to the estimated >800 G-protein coupled receptors in mammals, not a single

Mark P. Running (ed.), *G Protein-Coupled Receptor Signaling in Plants: Methods and Protocols*, Methods in Molecular Biology, vol. 1043, DOI 10.1007/978-1-62703-532-3_1, © Springer Science+Business Media, LLC 2013

classically defined GPCR has been unequivocally identified in plants. In part, this is due to the high burden of proof necessary to identify a protein as a classic GPCR. Classically defined GPCRs have exactly 7TMs, bind the Gα subunit of the heterotrimer, directly sense extracellular stimuli and transduce the extracellular signal into a structural change in Gα, which subsequently exchanges GDP for GTP and dissociates from the Gβγ dimer to effect downstream signaling. Notably, the first candidate GPCR in plants to be identified, GCR1, still remains a candidate GPCR by these criteria.

Although the burden of proof is extensive, the highest hurdle in identifying novel GPCRs in plants has been the intrinsically low sequence homology between GPCRs, which can be as low as 20 % even within a single metazoan GCPR family [4]. Not surprisingly, cross-kingdom BLAST queries using well-known metazoan GPCRs have not been substantially helpful. Thus, novel candidate GPCRs are commonly identified using bioinformatic methods that rely heavily on the conserved structural and biophysical characteristic of known GPCRs. The most common structural trait is the conserved topology comprising an extracellular N-terminus, 7TMs, and an intracellular C-terminus.

It is logical to use topology prediction as an initial bioinformatic screen, but topology prediction methods have their own inherent strengths and weaknesses [5]. To counter this problem, the output from multiple complementary methods, such as HMMTOP [6], TMHMM [7], and PHOBIUS [8, 9], are used to define a set of most probable 7TM proteins. Since the intrinsic hydrophobic structure of signal peptides can confound TM prediction, the additional function of signal peptide prediction provided within Phobius guides the topology assessment. The quasi-periodic feature classifier (QFC) is used to indirectly predict GPCRs based on conserved physiochemical properties [10], while the additional QFC ion channel filter removes unwanted contaminant sequences with similar characteristics. Combining the topological assessment with the physiochemical analysis yields an intermediate pool for subsequent analysis by GPCRHMM [11], which directly predicts GPCRs using distinctly different physiochemical and topological properties of GPCRs. Further refinement of the resulting candidate GPCR list can be made using classification strategies such as GPCRsIdentifier [12], or through the identification of conserved domains or motifs identified through the Pfam database or other resources. The final list of most-likely GPCRs is then tested for physical coupling to Gα using the mating based yeast split-ubiquitin assay [13].

Some notes of caution must be mentioned regarding the interpretation of the bioinformatic and wet-bench analyses. Computational programs, even when used in a combinatorial fashion, cannot provide proof of topology or in vivo function as a GPCR. The true topology of a protein can only be determined through empirical analysis, for which the membrane based split-ubiquitin is ill-suited. Positive yeast growth is occasionally observed even in homodimerization assays with ubiquitin-fused termini predicted to localize to the opposite sides of the plasma membrane. And, plant candidate GPCRs coupling to Gα in the yeast system must be confirmed in their native system, i.e., in plant cells, using independent protein–protein interaction assays such as BiFC, split-luciferase, or FRET assays. The method presented here provides a targeted approach to quickly enter the experimental phase of in plant GPCR testing.

The ultimate goal of a protein-protein interaction screen is to identify functional, physiologically relevant interacting partners of the gene interest. Recent evidence from a number of biochemical, structural, and evolutionary analyses indicate plant Gα subunits have an intrinsically slow GTPase activity, which leaves them primarily in the activated state [14, 15], in contrast to metazoan Gα subunits. This has been construed as evidence that GPCRs are not present in plants [16], but we have long-preferred the literal definition of a GPCR, i.e., a receptor that interacts with a Gα subunit, without a priori assignment of biochemical function or physiological role. Plants could still literally have GPCRs.

1.1 Explanation of Yeast Strain/Construct Relationships

The yeast strains THY.AP4 and THY.AP5 (hereafter AP4 and AP5) are of different mating type, and are both auxotrophic for leucine and tryptophan. The pMETYCgate construct (hereafter Cub) contains the *LEU2* gene and is used to complement the leucine auxotrophy of AP4. The pNXgate and pXNgate (hereafter Nub) based constructs contain the *TRP1* gene and are used to complement the tryptophan auxotrophy of AP5. Mating of the two strains transformed with the appropriate construct will yield a diploid strain autotrophic for both leucine and tryptophan.

Both yeast strains are also auxotrophic for adenine and histidine, and the constructs do not carry any genes for complementation. However, AP4 has been engineered to contain nuclear copies of *ADE2* and *HIS3* driven by LexA. This forms the basis for selection while testing protein interaction in the growth assay. A physical interaction between the two test hybrid proteins leads to the presence of free A-LexA-VP16 (PLV), a synthetic transcription factor, which activates the LexA driven genes and allows for growth on minimal media.

The Cub LEU2 gene is driven by the methionine repressible MET25 promoter. The use of a repressible promoter allows for the detection of nonspecific interactions of Nub based fusion proteins with Cub fusions by downregulating the amount of Cub fusion proteins present in the cell. Growth at higher methionine concentrations indicates higher specificity of the Nub fusion protein for the Cub fusion protein.

2 Materials

1. A Windows based computer, or a computer capable of emulating the Windows environment (*see* **Note 1**).
2. A text editing program with regular expression capabilities, such as EditPad Pro, that can open and work with large data files.
3. Internet access or local copies of the TMHMM, Phobius, HMMTOP, QFC, and GPCRHMM software.
4. Spreadsheet or relational database software, e.g., Microsoft Excel or Access.
5. The suite of split-ubiquitin yeast two hybrid variant vectors: Nub_{wt}-X, Nub_G-X, X-Nub_{wt}, X-Nub_G, and X-Cub (*see* **Note 2**).
6. The Saccharomyces cerevisiae THY.AP4 and THY.AP5 haploid yeast strains (*see* **Note 2**).
7. YPD media and plates supplemented with 2 mg/L adenine sulfate.
8. Synthetic complete agar plates without leucine (SC–Leu), without tryptophan (SC–Trp), and without leucine and tryptophan (SC–Leu–Trp).
9. Synthetic dextrose minimal media agar plates without amino acid supplementation (SD).
10. YEASTMAKER Yeast Transformation System 2 kit (Clontech) or the equivalent solutions prepared in the laboratory following the information in the kit.
11. U-shaped CELLSTAR suspension culture plates (Greiner Bio-One #650180) (*see* **Note 3**).

3 Methods

3.1 Bioinformatic Analyses and Overview

1. Obtain a whole-proteome sequence file, or compile a list of protein sequences to be analyzed, and convert it into FASTA format.
2. Check the protein sequences for quality using a text editor to ensure the computational programs will not stall or abort during analysis (*see* **Note 4**).
3. Remove redundant protein sequences using BLASTClust or other reductive method to create an input file.
4. Submit the input FASTA file to Web server based or local copies of TMHMM, Phobius, HMMTOP, GPCRHMM, QFC, and other programs as desired (*see* **Note 5**).
5. Format the output into tab delimited files (*see* **Notes 6** and **7**) using Subheadings 3.2 to 3.6.

6. Import the data into spreadsheet software or relational database software (e.g., Microsoft Excel or Access) for further analysis.
7. Identify candidate GPCRs using defined criteria (*see* **Note 8**) and proceed to the yeast split-ubiquitin assays to assess in vivo coupling.

3.2 Extracting TM Data from TMHMM

1. Delete all of the text identifying the data columns (e.g., "len=") except the "topology=" identifier (*see* **Note 9**). Also delete any trailing lines below the last prediction.
2. Make a column for the TMHMM N-terminal location data after the TM data column and before the topology column.

```
find: Topology        replace: \tTopology
```

3. For 0TM proteins, change the newly made N-terminal location field and the Topology field to "none." For 1TM proteins, change the N-terminus data while preserving the TM location data.

```
find: \t0\t\tTopology=(o|i)$        replace: \t0\tnone\tnone
find: \t1\t\tTopology=(o|i)         replace: \t1\tnone\t\1
```

4. Fill the N-terminal location field, and remove the "topology=" identifier for both extracellular (o) and intracellular (i) N-terminus predictions.

```
find: \t(\d\d?)\t\tTopology=o        replace: \t\1\tOUT\to
find: \t(\d\d?)\t\tTopology=i        replace: \t\1\tIN\ti
```

5. Check the tabbed formatted file for inconsistencies in structure.
6. Add a header line at the top of the file identifying each column.

3.3 Extracting TM and Signal Peptide Data from Phobius

1. Delete all of the lines, including the header line, above the first protein prediction. Also delete trailing lines below the last prediction.
2. Convert the whitespaces which separate the data columns into tabs, and add an additional column for the N-terminus location data.

```
find: ^(.+?\b)\s.+?(\d\d?)\s\s(Y|0)\s        replace: \1\t\2\t\3\t\t
```

3. For all proteins with signal peptides, enter the N-terminus location prediction.

```
find: \t(\d\d?)\tY\t        replace: \t\1\tY\tOUT
```

4. For 0TM proteins without a signal peptide, change the N-terminus location and topology prediction to "none."

find: \t0\t0\t\t(o|i)$	*replace*: \t0\t0\tnone\tnone

5. For TM proteins without a signal peptide, fill the N-terminus location field for both extracellular (o) and intracellular (i) N-terminus predictions.

find: \t\to	*replace*: \tOUT\to
find: \t\ti	*replace*: \tIN\ti

6. Check the file for inconsistencies and add a header line at the top of the file identifying each column.

3.4 Extracting Data from HMMTOP

1. Delete the first five characters (>HP:) that start every prediction.
2. Convert the whitespaces after the protein lengths and protein identifiers to tabs.

find: ^(\d.+?)\s(.+?\b.+?)\s\s?	*replace*: \1\t\2\t

3. Swap the first two columns to put the protein identifier first.

find: ^(\d.+?)\t(.+?\b.+?)\t	*replace*: \2\t\1\t

4. For TM proteins, convert the whitespaces following the N-terminus location and TM predictions to tabs.

find: (IN|OUT)\s{0,4}([1-9]\d?)\s{0,4}	*replace*: \1\t\2\t

5. For 0TM proteins, change the N-terminus location to "none," and add a new topology prediction column specifying "none."

find: (IN|OUT)\s{0,4}0	*replace*: none\t0\tnone

6. Check the file for inconsistencies and add a header line at the top of the file identifying each column.

3.5 Extracting Data from GPCRHMM

1. Delete all of the lines, including the header line, above the first protein prediction.

2. Delete whitespaces from the end of each line and replace the invisible carriage return and new line character with a simple line feed.

find: \s{1,}$\r\n *replace*: \n

3. Replace all of the whitespaces separating the columns with a single tab.

find: \s{2,} *replace*: \t

4. For sequences too short to be analyzed, remove the "Too short sequence" phrase, add a local score column, and change the "no" prediction to "short."

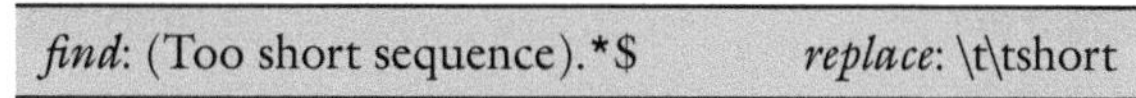

find: (Too short sequence).*$ *replace*: \t\tshort

5. Remove the hyphens from the local score column.

find: \t-\t *replace*: \t\t

6. Check the file for inconsistencies and add a header line at the top of the file identifying each column.

3.6 QFC and Mature Protein Topology

1. The QFC output is a tab delimited file ready for analysis, but the data labels should be changed to be consistent with the other output files.
2. Topological assessment of mature proteins by TMHMM and HMMTOP is performed after in silico removal of the signal peptide predicted by Phobius.

3.7 Cloning and Preparing Genes for Recombination

1. Choose suitable genes for the in vivo analysis using the output from the bioinformatic screen (*see* **Note 10**).
2. Design gene specific forward primers and gene specific reverse primers adapted with the SUS B1 and B2 linker sequences, respectively (*see* **Note 11**).
3. Amplify the candidate GPCR and gel-purify the product.
4. Clone the product into a suitable vector that will allow retrieval of the insert using restriction enzymes flanking the SUS adapted insert (*see* **Note 12**).
5. Verify the sequence of the clone (*see* **Note 13**).

6. Retrieve the B1 and B2 adapted inserts from the holding vector by restriction digest using sites that flank the insert yet do not cleave it (*see* **Note 14**).
7. Gel-purify the insert and quantify the concentration.

3.8 Preparation of SUS Vectors

1. Digest the X-Cub vector with *Pst*I and *Hind*III, and gel-purify the vector backbone away from the released ~0.8 kb fragment.
2. Digest the four Nub vectors with *Eco*RI and *Sma*I, and gel-purify the vector backbone away from the released ~1.8 kb fragment.
3. Quantify the concentration of the gel purified vectors (*see* **Note 15**).

3.9 In Vivo Recombination of Vector and Inserts

1. Prepare competent yeast cells using the standard efficiency protocol provided in the YEASTMAKER Yeast Transformation System 2 kit (Clontech).
2. Co-transform 100–120 ng of a cut Nub vector and 100 ng of a purified insert into competent AP5 cells. Perform all four Nub co-transformations separately (*see* **Note 16**).
3. Co-transform 100–120 ng cut X-Cub vector and 100 ng of the same purified insert into competent AP4 cells (*see* **Note 16**).
4. Plate co-transformed AP5 (Nub + insert) cells onto SC media–Leu, and plate co-transformed AP4 (Cub + insert) cells on SC media–Trp. Plate two dilutions and grow at 28–30 °C for 2–3 days.
5. Verify the identity of several colonies by colony PCR before proceeding further (*see* **Note 17**).

3.10 Yeast Mating

1. Prepare a list of matings to be performed (*see* **Note 18**).
2. Prepare YPDA agar plates to use as a mating surface (*see* **Note 19**).
3. Use a sterile toothpick to pick up cells from 7 to 8 colonies and resuspend into 0.5 ml YPD. Add 50–200 μl of the resuspended cells dropwise into a 5 ml YPD culture tube until it is just slightly turbid and grow overnight at 28 °C.
4. Collect cells from 1.5 ml of overnight culture using a benchtop centrifuge and discard the supernatant.
5. Resuspend cell pellet in 250 μl YPD.
6. For each mating pair, mix 15 μl of each suspension in a new tube (*see* **Note 20**).
7. Drop 5–8 μl of the mixture onto the semidry YPD plate making sure to leave ample room between the spreading droplets (*see* **Note 21**). Allow the droplets to dry before moving the plates.

8. Incubate the plates for 6–8 h at 28 °C for mating. Shorter times are not recommended.
9. Use a sterile flat tip toothpick to retrieve a portion of the mating spot and streak onto plates of SC–Trp–Leu agar plates.
10. Grow 2–3 days at 28 °C to select diploid colonies that contain the Nub and Cub expression vectors.

3.11 Growth Based Interaction Assays

1. Prepare a schematic diagram outlining sample organization on the plates (*see* **Note 22**).
2. Collect and mix 6–8 diploid colonies in a clean area of the selection plate.
3. Use the mixed colony paste as an inoculum to patch or streak the three assay plates (SD minimal media, SD + 200 μM methionine, and SD + 1 mM methionine) (*see* **Note 23**).
4. Incubate at 30 °C, and score daily for yeast growth beginning on day 2.
5. Positive interactions are usually evident on day 2 or 3, but can sometimes take up to 5 days.
6. Repeat the experiment, starting at the co-transformation step, to confirm positive interactions (*see* **Notes 24** and **25**).

4 Notes

1. Alternative operating systems and software may be required to locally run some programs. For example, a local copy of TMHMM2 runs on the UNIX platform and requires Perl.
2. The yeast strains and vectors are also available from the Arabidopsis Biological Resource Center (ABRC) [17] through The Arabidopsis Information Resource (TAIR) [18].
3. U-shaped wells work much better for mixing and retrieving small volumes than flat bottom culture plates. PCR strip tubes can also be used but they do not provide the same stability as plates.
4. Proteins should start with an "M," be completely on one line, have white spaces and inadvertent line breaks removed, and consist only of characters for the standard 20 amino acids and "X" for unknown. Some programs will not analyze sequences below or above a certain number of characters so these should be removed from the input FASTA files.
5. Use a block of the file if the number of sequences or total size of the file is above the server or program limit.
6. Several of the formatting steps could be condensed or coded more concisely, but optimizing the regular expression queries

makes the process opaque to novice users. These steps also increase comprehension and troubleshooting.

7. Correct format is essential. Data labels should be standardized across all of the output files and data types should not be mixed within a column. For example, if the column header for the protein identifier is called "ProtID" in the first formatted file the following files should have the same label. And, mixing numerical data and text data types within the same column will prevent correct import and sorting in subsequent analytical steps.
8. The most stringent criteria for identifying a candidate GPCR is a consensus 7TM topology and indirect and direct prediction as a GPCR. Alternative criteria could be made by weighting the prediction outputs unequally or including results from other complementary programs.
9. This can be done by finding the target text and replacing with nothing, e.g., if performed in Microsoft Word the "replace" field in the dialog box would be empty.
10. Also include Gα, and other proteins if desired, in the following cloning processes.
11. These relatively long primers will need additional processing beyond standard desalting, such as PAGE purification, to ensure oligonucleotide quality.
12. We typically use the BluntII TOPO vector from Invitrogen. An alternative vector should be used if the insert length is close to the linearized vector length to avoid difficulty at the gel separation and purification step.
13. Infrequently, clone sequence errors that originate from primer synthesis will appear near the tail of the B2 linker. In this case, sequence additionally 1–2 clones.
14. The recombination step is robust, and the presence of even a few hundred bases of vector derived sequence on each side does not inhibit success.
15. Subsequent quality analysis via agarose gel electrophoresis will also show that the purified linear X-Cub has greater apparent molecular weight than the four purified linear Nub vectors.
16. Co-transformation is performed using the protocol from the same YEASTMAKER Yeast Transformation System 2 kit. The components for the lithium acetate transformation method can be easily made in the lab.
17. The colony PCR step acts as a quality control step when testing a large number of candidate GPCRs. Sequencing can confirm the recombination has occurred correctly, but we have never identified a faulty recombination event.

18. Matings are performed 1:1 (Nub:Cub), but note that one set of directional matings requires four volumes of the Cub containing AP4 suspension to pair with the four unique Nub containing AP5 suspensions.
19. Air-dry the plates sufficiently to remove residual moisture to avoid excessive drying time after spotting the mating suspension.
20. This can be conveniently done in 96 well U-shaped CELLSTAR suspension culture plates (Greiner Bio-One #650180). Flat bottom culture plates make retrieval more difficult. PCR strip-tubes work well but do not offer the stability of rigid plates.
21. Mating occurs efficiently when a solid support such as the agar plate surface is used. Mating in liquid media is possible but requires additional optimization.
22. Maintaining a standard scheme will help expedite scoring. Keep test and control assays on the same plate for direct comparison; moist plates will support relatively faster yeast growth than drier plates.
23. If patching, the same toothpick can be used for all three plates as long as the lower methionine concentration plates are patched first. Take care to ensure patches within a plate and between plates are of equal density. Patching heavily so the patched cells are slightly visible does not cause spurious growth in negative assays, but excessive amounts or uneven patching will become visible in pictures. Spotting suspensions of the diploid cells is also possible.
24. Starting at later time points, e.g., post-mating, will not catch an erroneous mating or other problem after the identity of the haploid clone becomes "known" at the colony PCR stage.
25. The colorometric X-gal assay is sometimes performed as a semi-independent interaction assay, but in our hands we have never observed a case where the yeast growth assay was positive and the less sensitive X-gal assay was negative. Notably, the promotor driving auxotrophy complementation is the same promotor used to drive β-galactosidase production, and yeast growth is only supported in the presence of a sustained, sufficiently strong interaction while even weak/non-specific interactions can lead to the accumulation of β-galactosidase.

References

1. Wettschureck N, Offermanns S (2005) Mammalian G proteins and their cell type specific functions. Physiol Rev 85:1159–1204
2. Gookin TE, Kim J, Assmann SM (2008) Whole proteome identification of plant candidate G-protein coupled receptors in Arabidopsis, rice, and poplar: computational prediction and in-vivo protein coupling. Genome Biol 9:R120
3. Assmann SM (2002) Heterotrimeric and unconventional GTP binding proteins in plant cell signaling. Plant Cell 14(Suppl):S355–373
4. Horn F, Bettler E, Oliveira L, Campagne F, Cohen FE, Vriend G (2003) GPCRDB information system for G protein-coupled receptors. Nucleic Acids Res 31:294–297
5. Cuthbertson JM, Doyle DA, Sansom MS (2005) Transmembrane helix prediction: a comparative evaluation and analysis. Protein Eng Des Sel 18:295–308
6. Tusnady GE, Simon I (2001) The HMMTOP transmembrane topology prediction server. Bioinformatics 17:849–850
7. Krogh A, Larsson B, von Heijne G, Sonnhammer EL (2001) Predicting transmembrane protein topology with a hidden Markov model: application to complete genomes. J Mol Biol 305:567–580
8. Käll L, Krogh A, Sonnhammer EL (2004) A combined transmembrane topology and signal peptide prediction method. J Mol Biol 338: 1027–1036
9. Käll L, Krogh A, Sonnhammer EL (2007) Advantages of combined transmembrane topology and signal peptide prediction–the Phobius web server. Nucleic Acids Res 35:W429–432
10. Kim J, Moriyama EN, Warr CG, Clyne PJ, Carlson JR (2000) Identification of novel multi-transmembrane proteins from genomic databases using quasi-periodic structural properties. Bioinformatics 16:767–775
11. Wistrand M, Kall L, Sonnhammer EL (2006) A general model of G protein-coupled receptor sequences and its application to detect remote homologs. Protein Sci 15:509–521
12. Gao QB, Wang ZZ (2006) Classification of G-protein coupled receptors at four levels. Protein Eng Des Sel 19:511–516
13. Obrdlik P, El-Bakkoury M, Hamacher T, Cappellaro C, Vilarino C, Fleischer C, Ellerbrok H, Kamuzinzi R, Ledent V, Blaudez D, Sanders D, Revuelta JL, Boles E, Andre B, Frommer WB (2004) K^+ channel interactions detected by a genetic system optimized for systematic studies of membrane protein interactions. Proc Natl Acad Sci USA 101: 12242–12247
14. Jones JC, Duffy JW, Machius M, Temple BR, Dohlman HG, Jones AM (2011) The crystal structure of a self-activating G protein alpha subunit reveals its distinct mechanism of signal initiation. Sci Signal 4, ra8
15. Bradford W, Buckholz A, Morton J, Price C, Jones AM, Urano D (2103) Eukaryotic G protein signaling evolved to require G protein-coupled receptors for activation. Sci Signal 6, ra37
16. Urano D, Jones AM (2013) "Round up the usual suspects": a comment on nonexistent plant G protein-coupled receptors. Plant Physiol 161:1097–1102
17. The Arabidopsis Biological Resource Center. http://abrc.osu.edu/
18. The Arabidopsis Information Resource (TAIR). http://www.arabidopsis.org/

Chapter 2

Measurement of GTP-Binding and GTPase Activity of Heterotrimeric Gα Proteins

Swarup Roy Choudhury, Corey S. Westfall, Dieter Hackenberg, and Sona Pandey

Abstract

Heterotrimeric G-proteins are important signaling intermediates in all eukaryotes. These proteins link signal perception by a cell surface localized receptor to the downstream effectors of a given signaling pathways. The minimal core of the heterotrimeric G-protein complex consists of Gα, Gβ, and Gγ subunits, the G protein coupled receptor (GPCR) and the regulator of G-protein signaling (RGS) proteins. Signal transduction by heterotrimeric G-proteins is controlled by the distinct biochemical activities of Gα protein, which binds and hydrolyses GTP. Evaluation of the rate of GTP binding, the rate of GTP hydrolysis, and the rate of GTP/GDP exchange on Gα protein are required to better understand the mechanistic aspects of heterotrimeric G-protein signaling, which remains significantly limited for the plant G-proteins. Here we describe the optimized methods for measurement of the distinct biochemical activities of the Arabidopsis Gα protein.

Key words Heterotrimeric G-protein, Gα protein, GTP-binding, GTP hydrolysis, GTPase activity, GDP/GTP exchange, Arabidopsis GPA1

1 Introduction

G-proteins, localized at the inner surface of the plasma membrane, are key signaling intermediates in eukaryotes [1, 2]. In mammals, the importance of G-proteins in regulating fundamental signaling pathways involved in sensory perception (vision, olfaction, and taste), neurotransmission, hormone perception, and immunity-related cues has prompted in-depth characterization [1–3]. Such studies have revealed an elegant signaling mechanism, where the Gα subunit acts as a bimodal molecular switch, alternating between signal-dependent Gα·GDP and Gα·GTP conformations. In the inactive state, Gα·GDP associates with the Gβγ dimer, which represents the "off" signaling status. Signal perception by a G-protein

Swarup Roy Choudhury, Corey S. Westfall and Dieter Hackenberg have contributed equally to this work.

Mark P. Running (ed.), *G Protein-Coupled Receptor Signaling in Plants: Methods and Protocols*, Methods in Molecular Biology, vol. 1043, DOI 10.1007/978-1-62703-532-3_2, © Springer Science+Business Media, LLC 2013

coupled receptor (GPCR) causes the exchange of GTP for GDP on Gα. The GPCR thus acts as a guanine nucleotide exchange factor (GEF) causing the dissociation of the Gαβγ trimer. The active Gα·GTP and the freed Gβγ dimer then transduce the signal downstream by interaction with different effectors. The intrinsic GTPase-activity of Gα causes hydrolysis of the bound GTP, regenerating Gα·GDP, which re-associates with the Gβγ dimer [1, 2]. The GTPase activity of the Gα proteins is regulated by a group of accessory proteins, e.g., the regulators of G-protein signaling (RGS) proteins. The RGS proteins, by enhancing the GTPase activity of Gα, accelerate the rate of G-protein cycle [4, 5]. This classic signaling mechanism thus entails three biochemically distinct reactions of the Gα of the heterotrimer: *the rate of GTP binding, the rate of GTP hydrolysis, and the rate of GTP/GDP exchange,* which control all physiological responses regulated by G-proteins.

The biochemical activities of mammalian Gα proteins have been characterized in great detail, and it has been established that the rate of GTP-binding on Gα proteins is the rate limiting step for mammalian G-protein signaling. In contrast, only few plant Gα proteins have been characterized at the biochemical level, and many of them exhibit relatively higher rates of GTP-binding but slower GTPase activities [6–11]. This suggests that the GTPase activity of Gα could be the rate limiting step of the G-protein cycle in higher plants, although detailed characterization of multiple Gα proteins from different plant lineages is needed to make a generalized statement. In the following sections, we describe optimized methods for purification of plant Gα proteins and their biochemical characterization using radiolabeled or fluorescently labeled tags.

2 Materials

2.1 Reagents

1. Purified Gα protein (at least 95 % pure).
2. Purified recombinant RGS domain (for performing GTPase assay in the presence of RGS protein).
3. 5 mM GTP (Guanosine triphosphate).
4. 5 mM GDP (Guanosine diphosphate).
5. 5 mM ATP (Adenosine triphosphate).
6. 5 mM ADP (Adenosine diphosphate).
7. [^{35}S]GTPγS (Perkin-Elmer, 1,250 Ci/mmol).
8. [^{35}S]GTPγS-binding reaction buffer (Tris pH 8.0, 50 mM; $MgCl_2$, 10 mM; DTT, 1 mM).
9. [^{35}S]GTPγS-binding wash buffer (Tris pH 8.0, 20 mM; $MgCl_2$, 25 mM; NaCl,100 mM).
10. GF/B filter disks (Millipore).
11. Scintillation cocktail (OptiPhase HiSafe 2, Perkin-Elmer).

12. [α^{32}P]GTP (Perkin-Elmer, 3,000 Ci/mmol).
13. PEI cellulose TLC plates (Sigma-Aldrich).
14. TLC assay reaction buffer (Tris pH 8.0, 50 mM; $MgCl_2$, 10 mM; EDTA, 1 mM; DTT, 1 mM).
15. TLC assay stop solution (EDTA, pH 8.0, 0.5 M).
16. TLC assay developer solution (KH_2PO_4, pH 3.4, 0.5 M).
17. BODIPY-assay reaction buffer (Tris pH 8.0, 10 mM; $MgCl_2$, 10 mM).
18. 5 mM BODIPY-GTPγS (Invitrogen).
19. 5 mM BODIPY-GTP FL (Invitrogen).
20. Bovine serum albumin (fraction V powder, Sigma-Aldrich).
21. 5 mM Mant-GTP (Invitrogen).
22. 5 mM Mant-GDP (Invitrogen).
23. Mant reaction Buffer (Tris pH 8.0, 20 mM; NaCl, 100 mM; $MgCl_2$, 10 mM).

2.2 Equipment

1. Water bath (30 °C).
2. Vacuum-filtration device.
3. Scintillation vials.
4. Scintillation counter.
5. Cassette and phosphor screen.
6. Phosphorimager.
7. 96 Well Polystyrene Microplates (Greiner Bio One).
8. Fluorescence microplate reader (FLUOstar Optima, BMG Lab Technologies or equivalent).
9. Amicon© Centrifugal Filters (10 KDa cut off).
10. Olis© DM45 spectrofluorimeter with a 150-W xenon lamp, stopped-flow accessory, and water bath.
11. 420 nm cutoff-filter.
12. Fitting software such as KinTek Global Kinetic Explorer [12].

3 Methods

3.1 Radiolabeled GTP-Binding Assay

1. Adjust purified, recombinant Gα protein concentration to 100 nM in 200 μL of [^{35}S]GTPγS-binding reaction buffer.
2. Add 0.2 μM [^{35}S]GTPγS and incubate samples at 30 °C in a water bath for up to 2 h (*see* **Note 1**).
3. Take a small reaction aliquot (5–10 μL) in a separate tube kept chilled on ice at desired time points (*see* **Note 2**).
4. Add 1 mL of ice-cold wash buffer to the aliquot immediately to stop the reaction. Keep tubes on ice.

5. Wash a GF/B filter with ice cold wash buffer using a vacuum filtration device.
6. Filter the reaction mixture through the GF/B filter, followed by washing with 3–5 mL of ice cold wash buffer.
7. Dry GF/B filters (air dry or oven dry) and place them in a scintillation vial.
8. Add 5 mL of scintillation cocktail to each vial and measure the incorporated radioactivity using a scintillation counter.
9. Analyze and fit data using appropriate software of your choice.

3.2 GTPase Assay Using Thin Layer Chromatography

1. Adjust purified, recombinant Gα protein concentration to 100 nM in 200 μL of TLC assay reaction buffer.
2. Add 20 pmol [α^{32}P]GTP and incubate samples at 30 °C in a water bath for up to 2 h.
3. Take a 10 μL aliquot in a separate tube kept chilled on ice at desired time points (*see* **Notes 1** and **3**).
4. Add 10 μL of 0.5 M EDTA to the tubes immediately to stop the reaction. Keep tubes on ice.
5. Spot 1 μL of reaction on to PEI-cellulose TLC plates and air dry. Repeat this step two more times, each time spotting on top of the original spot.
6. Develop plate in 0.5 M KH_2PO_4 (pH 3.4) solution and dry.
7. Expose plate to a phosphorimager screen for 6–12 h. Scan using a phosphorimager.

3.3 GTP Binding Assay Using BODIPY-GTP or BODIPY GTPγS

1. Use 50 mL BODIPY assay buffer to make 10 nM BODIPY-GTPγS-FL or BODIPY-GTP-FL solutions. BODIPY-GTPγS-FL is used for detection of GTP-binding only while BODIPY-GTP-FL can be used for estimation of both GTP-binding and GTP-hydrolysis.
2. Adjust Gα protein concentration to 200 nM in BODIPY assay buffer.
3. Add 100 μL of 200 nM Gα protein in 100 μL assay buffer in 96 well, flat bottom polystyrene microplates and mix properly. A total three to five replicates for each reaction are highly recommended. Reaction mixes containing 100 nM BSA and without any protein should be included as negative controls with each run.
4. Wash the injection tubes getting in and out of the fluorescent plate reader with sterile water, followed by BODIPY-assay buffer.
5. Add 100 μL of 10 nM BODIPY-GTPγS or 10 nM BODIPY-GTP FL solutions to the protein solution to initiate the reaction using the automatic injector of the plate reader. The final protein concentration is 100 nM.

6. Record fluorescent reading using following set up for FLUOstar Optima, BMG Lab Technologies fluorescent plate reader: Positioning delay-0.1 s, number of kinetic windows-1, number of cycles-100, number of flashes per cycle-10, filter and integration-fluorescence intensity, number of multichromatics-1, excitation filter-485 nm, emission filter-520, gain-1,500, pump speed-420 μL/s, pump prime volume-500 μL, temperature-25 °C, data recording-100 cycles every 16 s (total time per read: ~30 min). These parameters can be adjusted to desired specifications for any available fluorescent plate reader.
7. For competition experiments, 5 μM of non-labeled nucleotides (GTP/GDP/ATP/ADP) can be added to the assay buffer before starting the reaction. The additives should also be included in the negative controls.
8. For the effect of RGS protein on the rate of GTP hydrolysis, recombinant RGS domain can be added to the reaction mix at a stoichiometric ratio of RGS to Gα at 2:1. A reaction mix containing only RGS protein should be included as a negative control for these assays.
9. Subtract fluorescent readings for the negative controls at each time point and plot relative fluorescence of the proteins as a factor of time using any software of your choice. When using BODIPY GTP-FL, the increase in fluorescent with time denotes GTP-binding and subsequent decrease in fluorescent with time denotes GTP hydrolysis by Gα protein.

3.4 GTP-Binding Using Mant-GTP

1. Using the Amicon©, exchange purified protein into Mant-GTP reaction buffer and concentrate to 2 μM, keep on ice (*see* **Note 4**).
2. 5 mM Mant-GTP should be diluted into reaction buffer to form 20, 30, 40, 50, 100, 160 μM, keep on ice (*see* **Note 5**).
3. Wash fluorometer injection tubes (Olis© DM45 spectrofluorimeter or equivalent) 3 times with reaction buffer.
4. Fill stopped-flow syringes with Gα protein solution and Mant-GTP solution.
5. Run 2 injections to wash machine.
6. Let solutions come up to 20 °C for 10 min.
7. Set excitation wavelength to 280 nm and use the >420 nm cutoff for emission. This actually tracks the FRET between a tryptophan and the Mant group. If there is no FRET, an excitation of 350 nm can be used to directly measure Mant fluorescence. FRET is usually less noisy, so it is recommended if possible.
8. Run 1 injection, adjust PMT voltage as needed.
9. Run 9–11 injections, collecting around 1–2 s per injection.
10. Repeat **steps 3–7** for each Mant-GTP concentration.

3.5 Analysis of Mant-GTP Binding Data

1. Normalize all sample runs.
2. Average normalized samples.
3. Repeat for each Mant-GTP concentrations.
4. Depending on the protein, fit the data to either a single or double exponential curve.
5. A graph of Mant-GTP versus observed rate constants should be a line with a slope equal to the binding constant and an *x*-intercept of the dissociation constant. The Mant-GTP concentration used should be half of starting concentration due to the 1:1 mixing with the protein solution in the stopped-flow.
6. For data that is fit to double exponential curves, one set of observed rate constants will behave linearly with a slope equal to the binding constant and an *x*-intercept of the dissociation constant. The second observed rate constant should be constant for all Mant-GTP concentrations and corresponds to a conformational change upon binding [13] (*see* **Note 6**). The equation for the linear fit is $k_{obs} = k_{-1} + k_1[\text{Ligand}]$.

3.6 Measurement of GDP/GTP Exchange

1. Using the Amicon©, exchange purified protein into reaction buffer and concentrate to 2 μM, keep on ice as in Subheading 3.4 **step 1** (*see* **Note 4**).
2. Add Mant-GDP directly to the protein solution to a final concentration of 5 μM (*see* **Note 7**).
3. Make solutions of GTP (no Mant group) of 40, 80, 160, 320 μM in Mant reaction buffer.
4. Fill stopped-flow syringes with Gα protein plus Mant-GDP solution and GTP solution.
5. Run 2 injections to wash machine.
6. Let solutions come up to 20 °C for 10 min.
7. Set excitation wavelength to 280 nm and use the >420 nm cutoff for emission.
8. Run 1 injection, adjust PMT voltage.
9. Run 9–11 injections collecting 30–45 s per injection.
10. Repeat **steps 4–9** for each GTP concentration.

3.7 Analysis of Nucleotide Exchange Data

1. Normalize all 9–11 sample runs and average normalized samples.
2. Repeat for each GTP concentrations.
3. Data for each GTP concentration should be able to fit to a single-exponential curve. If not, removal of beginning points or end points might be necessary.
4. A graph of the observed rate constants should be hyperbolic. If the rate constants do not reach a plateau, higher GTP concentrations might need to be used.

5. The observed rate constant for saturated GTP corresponds to the dissociation rate of the Mant-GDP [14] (*see* **Note 8**).

4 Notes

1. Strictly follow all safety procedures for handling and disposal of radioisotopes and liquid and solid waste containing radiochemicals. Survey the usage area and equipment before and after the use of radioisotopes. Follow your institutional guidelines for recording of radioisotope usage and disposal.
2. The aliquots should be collected every 30 s to 1 min, if possible to achieve better resolution of binding kinetics at the initial time points. Three to five aliquots should be collected for each time point. For strong GTP-binding proteins, 30–60 min incubation is sufficient.
3. The aliquots should be collected every 30 s to 1 min at the initial time points and should be added directly to ice cold tubes containing chilled EDTA. Three to five aliquots should be collected for each time point. Maximum 18 samples can be spotted on a standard TLC plate for each run.
4. Protein concentration needs to be calculated using accurate techniques, such as absorbance at 280 nm.
5. Exact concentrations of protein, Mant-GTP, and Mant-GDP will depend on the protein being used. To aid in analysis, always maintain pseudo-first order kinetics by keeping the nucleotide 10× higher than the protein.
6. This analysis only works if you always maintain pseudo-first order kinetics. If you are not pseudo-first order, you must either fit computationally, using a program like KinTek©, or use a more complicated fitting equation [13].
7. Mant-GDP concentration needs to be high enough to saturate binding sites.
8. Some proteins might have a negative hyperbolic curve for the exchange reaction. *See* ref. 14 for this circumstance.

Acknowledgments

The research in corresponding author's laboratory was supported by a US Department of Agriculture/Agriculture and Food Research Initiative grant (2010-65116-20454) and a National Science Foundation grant (MCB-1157944). C.S.W. was supported by a US Department of Agriculture/Agriculture and Food Research Initiative predoctoral research fellowship (MOW-2010-05240).

References

1. Offermanns S (2003) G-proteins as transducers in transmembrane signalling. Prog Biophys Mol Biol 83(2):101–130
2. Cabrera-Vera TM et al (2003) Insights into G protein structure, function, and regulation. Endocr Rev 24(6):765–781
3. Oldham WM, Hamm HE (2008) Heterotrimeric G protein activation by G-protein-coupled receptors. Nat Rev Mol Cell Biol 9(1):60–71
4. Dohlman HG, Thorner J (1997) RGS proteins and signaling by heterotrimeric G proteins. J Biol Chem 272(7):3871–3874
5. Siderovski DP, Willard FS (2005) The GAPs, GEFs, and GDIs of heterotrimeric G-protein alpha subunits. Int J Biol Sci 1(2): 51–66
6. Jones JC, Temple BR, Jones AM, Dohlman HG (2011) Functional reconstitution of an atypical G protein heterotrimer and regulator of G protein signaling protein (RGS1) from *Arabidopsis thaliana*. J Biol Chem 286(15): 13143–13150
7. Jones JC et al (2011) The crystal structure of a self-activating G protein alpha subunit reveals its distinct mechanism of signal initiation. Sci Signal 4(159):ra8
8. Johnston CA et al (2007) GTPase acceleration as the rate-limiting step in Arabidopsis G protein-coupled sugar signaling. Proc Natl Acad Sci USA 104(44):17317–17322
9. Seo HS, Choi CH, Lee SY, Cho MJ, Bahk JD (1997) Biochemical characteristics of a rice (*Oryza sativa* L., IR36) G-protein alpha-subunit expressed in *Escherichia coli*. Biochem J 324(Pt 1):273–281
10. Bisht NC, Jez JM, & Pandey S (2011) An elaborate heterotrimeric G-protein family from soybean expands the diversity of plant G-protein networks. New phytol 190:35–48
11. Roy Choudhury S et al (2012) Two chimeric regulators of G-protein signaling (RGS) proteins differentially modulate soybean heterotrimeric G-protein cycle. J Biol Chem 287(21): 17870–17881
12. Johnson KA (2009) Fitting enzyme kinetic data with KinTek Global Kinetic Explorer. Methods Enzymol 467:601–626
13. Kozlov AG, Lohman TM (2002) Stopped-flow studies of the kinetics of single-stranded DNA binding and wrapping around the *Escherichia coli* SSB tetramer. Biochemistry 41(19): 6032–6044
14. Wu XM, Gutfreund H, Chock PB (1992) Kinetic method for differentiating mechanisms for ligand exchange reactions: application to test for substrate channeling in glycolysis. Biochemistry 31(7):2123–2128

Chapter 3

Biochemical Analysis of the Interaction Between Phospholipase Dα1 and GTP-Binding Protein α-Subunit from *Arabidopsis thaliana*

Jian Zhao and Xuemin Wang

Abstract

Phospholipase Ds (PLDs) play diverse roles in plant lipid metabolism and cellular signaling processes. The sole canonical G-protein α-subunit (Gα) in Arabidopsis also plays multiple roles in plant growth and cellular signaling processes. Interestingly, overlapping functions of PLD and Gα have been indicated in many cellular processes, including abscisic acid (ABA)-mediated stomata movement and water loss, gibberellic acid (GA)-regulated seed germination, and auxin signaling. This interaction between PLD and Gα has also been suggested in biochemical and physiological studies. Here we described the methods used for studying the interaction between the major PLD form PLDα1 and Gα. From pulldown experiments with purified bacterially expressed PLDα1 and Gα, co-immunoprecipitation of plant protein extract, and yeast two-hybrid assay, we showed that there is a physical interaction between PLDα1 and Gα, and identified a key DRY motif in PLDα1, which is an essential element for the interaction. The interaction of PLDα1 and Gα can be affected by factors like GTP or GDP, but it also affected PLD phospholipase activity and Gα GTPase activity in turn.

Key words Phospholipase D, G-protein α-subunit, Protein–protein interaction, Activity assay

1 Introduction

Phospholipase D (PLD) hydrolyzes phospholipids into phosphatidic acid and various head groups, depending on types of phospholipid substrates [1]. Many PLDs have been characterized from many plant species since the cloning of the first eukaryotic PLD gene in castor bean [2]; there are at least 12 PLD isoforms with different properties in Arabidopsis, and they play diverse roles in lipid metabolism and cellular regulation [1]. So far, the biological functions of plant PLDs have been extended to hormone signaling (abscisic acid, jasmonate, auxin, and gibberellic acid), environmental stress responses (drought, freezing, wounding, heavy metal, and phosphorus nutrition), vesicle trafficking and cytoskeleton dynamics, and disease responses (*see* reviews [3, 4]).

Mark P. Running (ed.), *G Protein-Coupled Receptor Signaling in Plants: Methods and Protocols*, Methods in Molecular Biology, vol. 1043, DOI 10.1007/978-1-62703-532-3_3, © Springer Science+Business Media, LLC 2013

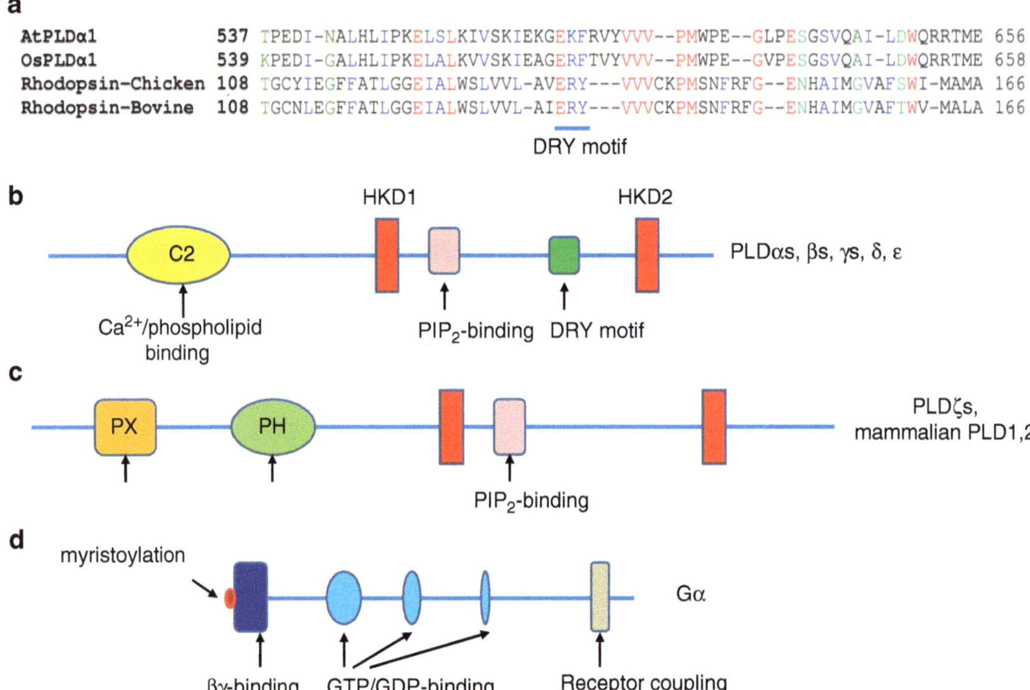

Fig. 1 DRY motif sequence alignment and schematic structures of PLDs and Gα. (**a**) Rice PLD α1 (OsPLD α1, accession number Q43007), Arabidopsis PLDα1(AtPLDα1, accession number BAB02304) Rhodopsin from chicken (Swiss-Prot accession number P22328), Rhodopsin from bovine (accession number NP_001014890) are used for alignment of DRY motif sequences. (**b**) Domain structures of the C2-PLD subfamily. The C2 domain is for Ca^{2+}/phospholipid binding and the DRY motif for G-protein interaction. (**c**) Domain structures of the PX/PH-PLD subfamily. The phox (PX) and pleckstrin homology (PH) domains have PIP_2 binding and regulatory functions. All PLDs contain the conserved duplicated HKD1 and two motifs that are involved in catalysis. This is adapted from Wang [3]. (**d**) Gα structure has conserved myristoylation and βγ-binding sites in N-terminus for attaching to the membrane and binding with Gβ and Gγ subunits, respectively, three GTP binding domains, and receptor coupling site in C-terminus

PLDα1 produces phosphatic acid under several stress conditions and has a multifaceted function [4]. A role in regulating plant PLD has been proposed for G-proteins in several physiological, biochemical, and genetic studies [5]. In mammalian cells, PLDs physically interact with different types of small G-proteins, such as RhoA, ADP-ribosylation factor, dynamin GTPase, and GTPase Rac2, and regulates activity through a G-protein coupled receptor (GPCR) signaling [6–9]. A previous study shows that PLDα1 binds to Gα through a motif analogous to the DRY motif present in mammalian GPCRs, and that this binding modulates the activity of PLDα1 and Gα [10]. Further analysis verified the high similarity in primary sequences for DRY motives between plant PLD α1 and rhodopsins (Fig. 1). The rice PLDα1 has even more significant DRY motif signature than AtPLDα1 does. The interaction may take place in a membrane-associated manner, according to their secondary structure predictions and membrane-association properties under certain circumstances. The biological significance of the interaction of PLDα1 and Gα has been indicated in ABA-mediated stomata movement regulation [11].

According to protein structure, PLDs can be classified into two distinctive subfamilies: the C2-PLDs, including PLDα1 and most other plant PLDs, and the PX/PH-PLDs, including Arabidopsis PLDζ1 and ζ2 and mammalian PLDs (Fig. 1; ref. 3). The two subfamilies all have two duplicated HKD and PIP_2 binding motifs within the C-terminus as their common features [3, 4]. The DRY motif in PLDα1 C-terminus responsible for interacting with Gα is very similar to that in mammalian orphan GPCRs (Fig. 1, refs. 6, 10). Similar to the way mammalian PLDs interact with and activate small GTPases, Arabidopsis PLDα1 seems also to act as a GAP (GTPase activating protein) when interacting with Gα [10]. However, mammalian PLDs interact with GTPases at different domains and sequences in their structures. For example, the PX domain in mammalian PLDs binds phosphoinositide lipids, the small G-protein Rac2, and dynamin GTPase [7, 9, 12]. The C-terminus of mammalian PLDs contains the catalytic regions, HKD domains which are critical for PLD activity and are required for interaction with small GTPase Rho A and Rac1 [8, 9, 12].

In Arabidopsis, there is one canonical Gα subunit, one Gβ subunit, AGβ1, and two Gγ subunits, AGG1and AGG2. Recently, some components and interactions of Arabidopsis heterotrimeric G protein complex have been revealed [13–15]. Both Gα and Gγ1 or Gγ2 may be associated with the plasma membrane, depending on their two putative lipidation motifs [13]. Studies using various technologies, including fluorescence resonance energy transfer (FRET), florescent protein fusion, and gel filtration, indicate that interactions between cytosolic AGβ1 and plasma membrane-localized AGG1 or AGG2 are independent of Gα, and direct association between AGG1 or AGG2 with the Gα subunit is also detected. Gα1-AGβ1-AGG1 heterotrimers at the plasma membrane are also associated with a large protein complex of 400–700 kDa, whereas Gα itself is also associated with smaller complexes in the 200–400 kDa range [14]. Activation of Gα causes dissociation of Gα1-AGβ1-AGG1 heterotrimers complex [14]. Thus, Gα can be associated with the plasma membrane in a protein complex with AGβ1-AGG1 or a complex with unknown proteins.

Efforts have been made to reveal the G-protein complex in order to uncover G-protein activation/inactivation and biological functions. In addition to PLDα1, several other proteins have been shown to interact with Gα in Arabidopsis, and those include GPCR-like proteins GCR1, GTG1, and GTG2 [16, 17], a RGS (regulator of G protein signaling) AtRGS1 [18], a plastid protein THYLAKOID FORMATION1 [19], a cupin domain protein

AtPirin1 [20], and a prephenate dehydratase [21]. Among these Gα-interacting proteins, AtRGS1 accelerates Gα GTPase activity and stabilizes the GDP-bound, heterotrimeric complex with AtRGS1 [18]; GTG1 and GTG2 also have intrinsic GTP-binding and GTPase activity that is inhibited by Gα binding [17]; PLDα1-Gα interactions inhibits PLDα1 activity but stimulates Gα GTPase activity [10], indicating that PLDα1 acts like a G-protein activator. Interestingly, most biological functions of Gα-interacting proteins, such as PLDα1, AtPirin1, AtRGS1, GCR1, GTG1, and GTG2, are related to ABA signaling. However, details about heterotrimeric G-protein complexes in planta and the dynamic changes and biological functions of the complex remain elusive.

2 Materials

2.1 Basic Molecular Biology Reagents and Equipments

1. Agarose gel and polyacrylamide gel electrophoresis reagents (Gibco/BRL), western blotting materials, polyvinylidene difluoride (PVDF), thermocycler, restriction endonucleases.
2. Various buffers:Tris–HCl, Phosphate-buffered Saline (PBS, 10×, 1.37 M NaCl, 27 mM KCl, 100 mM Na_2HPO_4, 18 mM KH_2PO_4, pH 7.4), Tris-buffered Saline with Tween (TBS-T, 10×, 1.37 M NaCl, 27 mM KCl, 250 mM Tris–HCl, pH 7.4, 1 % Tween-20), Running buffer (10×, 250 mM Tris, 1.92 M glycine, 1 % (w/v) SDS) and transfer buffer (10×, 250 mM Tris, 1.92 M glycine, 0.5 % (w/v) SDS, 20 % methanol).
3. Bacteria and yeast growth media, antibiotics (Gibco/BRL).
4. PLDα1 polyclonal antibodies raised in a rabbit against the 13 C-terminal amino acid residues as described previously [10, 22].
5. Gα polyclonal antibody raised in a rabbit against a synthetic oligopeptide, DETLRRRWLLFAGLL, corresponding to the C terminus conserved sequence of Gα [23], as a generous gift of Dr. Hong Ma in Pennsylvania State University [24].
6. Alkaline phosphatase conjugated second goat polyclonal antibody against rabbit immunoglobulin (Sigma).

2.2 Cloning and Site-Directed Mutagenesis

1. Arabidopsis PLDα1 cDNA in pBlue SK [22]; Gα cDNA clone [24].
2. 6× His-tagged fusion purification kit (Promega) and GST-fusion protein purification kit (Promega). QuikChange XL site-directed mutagenesis kit (Stratagene, CA).
3. Primers: a forward primer PLDα15, 5'-GCGGATCCATGGC GCAGCATCTGTTGCACG-3' (*Bam H1* site underlined) and

a reverse primer PLDα13, 5'-CGGAGCTCTTAGGTTGTA AGGATTGGAGGC-3' (*Sac 1* site underlined) for subcloning PLDα1 cDNA into pET28-(+)a for 6× His-tagged fusion.

4. A forward primer Gα5 5'-GAATTCATGGGCTTACT CTGCAGTAGAA-3' (*Eco R1* site underlined) and a reverse primer Gα3, 5'-CTCGAGTCATAAAAGGCCAGCCTCC AGTA-3' (*Xho 1* site underlined) for subcloning Gα cDNA into pGEM T-easy vector and subsequently pGEX-4T vector for glutathione S-transferase (GST) fused protein and pJG4-5 for yeast two-hybrid assay.
5. Three complementary reverse primers for mutating E563A, K564A, and F565A in PLDα1 as follows: E563A, 5'-GATTG AGAAAGGA**GCG**AAGTTCAGGGTCTATGTTGTGG-3α; K564A, 5'-GATTGAGAAAGGAGAG**GCG**TTCAGGGTCT ATGTTGTGG-3α; and F565A, 5'-GAGAAAGGAGAGAAG **GCC**AGGGTCTATGTTGTGG-3'.

2.3 Functional Expression of PLDα1 and Gα

1. *Escherichia coli* strain DH5α, BL21(DE3).
2. Cloning and expression systems: pGEM T-easy cloning (Promega USA). His- and GST-tagged protein expression vectors pET28(+)a and pGEX-4T, respectively.
3. IPTG (Isopropyl-l-thio-α-D-galactopyranoside), PMSF (phenylmethanesulfonyl fluoride), Glutathione Sepharose beads (Pharmacia), Ni-affinity agarose beads (Promega). Protease inhibitors (5 μg each of aprotinin, leupeptin, and antipain) from Sigma.

2.4 Yeast Two-Hybrid Assay

1. Yeast strain *Saccharomyces cerevisiae* EGY48 (MATα *trp1 his3 ura3 leu2*::6 LexAop-LEU2) containing LacZ reporter gene plasmid pSH18-34 (*URA3*, 2 μ, ApR, 8 ops.-lacZ).
2. Yeast two hybrid vectors pEG 202 (constitutively ADH promoter-driven LexA expression, followed by a polylinker for making the bait fusion protein, His selectable marker, 2 μ) and pJG 4–5 (GAL1 promoter-driven B42-HA tag followed by a polylinker for target fusion protein, Trp selectable marker, 2 μ).

2.5 PLDα1 and Gα activity Assay

1. Phospholipid substrates: L-phosphatidylcholine (PC), phosphatidylethanolamine, ^{3}H-labeled 1,2-dipalmitoyl-3-phosphatidyl-[*methyl*-^{3}H]choline (3.636×10^{-7} μmol/dpm, TRK673, Amersham).
2. EnzChek® phosphate assay kit (Molecular Probes, Eugene, OR).

2.6 Effects of GDP/GTP on PLDα1-Gα Interaction

1. Analytic grade of GTP, GDP, GDPαS, GTPαS, and Gpp(NH)P from Sigma.

3 Methods

3.1 Functional Expression of PLDα1 and Gα in *Escherichia coli*

3.1.1 For PLDα1 expression

1. Subclone 2.4-kb *Arabidopsis* PLDα1 cDNA from pBlue SK into the pGEM T-easy vector. Then the PLDα1 cDNA insert is digested with *Bam H1* and *Sac1*, then ligate the digested fragment into the pET28(+)a vector, and transform *E. coli* BL21(DE3) with the sequenced correct construct to produce PLDα1 with six histidine residues fused at the N terminus.
2. The transformed *E. coli* BL21(DE3) are grown to OD_{600nm} 0.6–0.8 in 37 °C shaker. IPTG (Isopropyl-1-thio-α-D-galactopyranoside, 0.1 mM final concentration) is added to induce PLDα1 expression at room temperature for 12 h (*see* **Note 1**).
3. Precipitate the bacteria pellets, resuspend them in phosphate-buffered saline plus 2 mM PMSF (phenylmethanesulfonyl fluoride), and lyse by sonication. Then centrifuge the lysate at 12,000 × *g* for 10 min and transfer the supernatant into a fresh tube, to which Ni-affinity agarose beads are added and mixed. Incubate the mixture at rolling shaker at 4 °C for 1 h.
4. Centrifuge the incubation beads at 500 × *g* at 4 °C and wash the bead pellets with a washing buffer containing 20 mM Tris–HCl, 0.5 M NaCl, and 20 mM imidazole at pH 8.0 three times (*see* **Note 2**).
5. Elute 6× His-PLDα1 from Ni-affinity agarose beads with an elution buffer with 1 M imidazole, Tris–HCl, and 0.5 M NaCl (His-tagged PLDα1 agarose beads may also be used directly for assaying PLDα1 activity). The purified PLDα1 is measured by the Bradford method using a Bio-Rad kit with bovine serum albumin (BSA) as a standard. The protein is stored in 20 % glycerol at −80 °C when not in use.

3.1.2 For Gα expression in *E. coli*

1. Subclone the 1.3-kb *Arabidopsis* Gα cDNA from pGEM7Zf(+) [24] into the pGEM T-easy vector firstly at *Eco R1* and *Xho 1* sites, and then into pGEX-4 T to produce Gα with glutathione *S*-transferase (GST) fused at the N terminus.
2. Transform *E. coli* BL21(DE3) with the recombinant plasmid to express the GST-Gα fusion and fee GST as control for following experiment (*see* **Note 3**).
3. Induce GST-Gα fusion expression in *E. coli* BL21(DE3) when bacteria grown at OD_{600nm} between 0.4 and 0.7 with IPTG (0.2 mM final concentration) at room temperature overnight.

4. Pellet cells at 5,000 ×*g* for 8 min, resuspend cells completely in ice cold lysis buffer 10 mM Tris–HCl (pH 7.0), 100 mM KCl, 0.1 mM EDTA, 0.25 % Triton X100, 2 mM PMSF (PMSF freshly prepared, *see* **Note 4**) and rupture cells by sonication.
5. Centrifuge the lysate at 10,000 ×*g* at 4 °C for 15 min. take the supernatants into a fresh tube and add freshly washed (with lyses buffer) glutathione Sepharose beads (the amount of Glutathione Sepharose beads may depend on the volume of bacteria cells). Incubate tube on end-over-end mixing shaker at 4 °C for 2 h.
6. Centrifuge the incubation mixture at 500 ×*g* for 2 min and remove the supernatant, wash beads with cold lyses buffer for five times and then remove most of the buffer.
7. GST-Gα glutathione Sepharose beads may be used for PLDα1 pulldown assay. For Gα GTPase activity assay, add elution buffer 100 mM Tris–HCl (pH7.0) containing 100 mM KCl and 25 mM glutathione to elute GST-Gα from beads by incubation at room temperature for 15 min and centrifugation at 1,000 rpm for 1 min. Transfer supernatant to a new tube, test protein concentration by the Bradford method.

3.2 Site-Directed Mutagenesis in PLDα1 DRY Motif

1. Use QuikChange XL site-directed mutagenesis kit to do the mutagenesis of the three codons in the DRY motif of PLDα1. Read the instruction carefully.
2. Perform PCR, using wild-type PLDα1 cDNA in pET28(+)a served as the PCR template, and E563A, K564A, and F565A forward and reverse primers, according to the kit instructions.
3. Verify the mutant codons by DNA sequencing, the corrected ones are transformed into BL21(DE3) for expression.

3.3 PLDα Pulldown by GST-Gα Beads

1. Add the purified GST-Gα-agarose beads (~0.05 μmol of protein, as determined with the Bradford method) or GST-agarose beads (as a control) to the plant protein extracts (PE); partially bacterially expressed PLDα1 or PLDα1 mutants (~0.025–0.15 μmol protein) in a microcentrifuge tube containing a pulldown buffer 50 mM Tris–HCl (pH 8.0), 200 mM NaCl, 2 mM $MgCl_2$, 5 mM $CaCl_2$, and protease inhibitors (5 μg each of aprotinin, leupeptin, and antipain) in a total volume of 150 μl.
2. Incubate the pulldown reaction microcentrifuge tube on an end-over-end mixing shaker at 4 °C for 3 h. Then beads are pelleted by centrifugation at 500 ×*g* for 3 min, and washed three times with the cool pulldown buffer containing 0.01 % Triton X-100 (stored at 4 °C). Pulldown beads are either

dissolved in 30 μl SDS-PAGE loading buffer and subsequently subjected to 10 % SDS-PAGE, or resuspended for PLDα1 activity assay.

3. Optional: the washed pulldown beads may also be washed with 50 μl of 20 mM reduced glutathione in 50 mM Tris–Cl (pH 8.0) to elute GST-Gα fusion or GST as well as any proteins bond to them from agarose beads, centrifuge the tubes at maximum speed for 2 min. The supernatants are used for immunoblotting detection or PLDα1 activity assay.
4. The SDS-PAGE gels are then followed by transferring proteins on gels to PVDF membrane and immunoblotting with antibodies specific to PLDα1. PLDα1 bands on blots are visualized by staining alkaline phosphatase activity conjugated to a second antibody (goat against rabbit immunoglobulin) (*see* **Notes 5** and **6**). The details for SDS-PAGE, protein blotting, and band detection are described previously [10].

3.4 Gα Immunoprecipitation by PLDα1 Antibody

1. Extract plant total proteins from fully developed leaves of *Arabidopsis thaliana* (Columbia). Frozen leaves with liquid N_2, homogenize with an extraction buffer 50 mM Tris–HCl (pH 7.5) containing 80 mM KCl, 2 mM EDTA, 5 mM dithiothreitol, and protease inhibitors (5 μg each of aprotinin, leupeptin, and antipain, 2 mM phenylmethanesulfonyl fluoride). Centrifuge the homogenate at 10,000 × *g* at 4 °C for 15 min, and take the supernatant as protein extracts for pulldown assay and co-immunoprecipitation (*see* **Note 7**).
2. Add 10 μl of PLDα1 antibody or preimmuno sera to 100 μl protein extracts (PE), 40 μl of Co-immunoprecipitation (CIP) buffer 100 mM Tris–HCl (pH 7.5) containing 500 mM KCl, 5 mM EDTA, 8 mM dithiothreitol, 0.4 % Triton X-100, and 5 mM phenylmethanesulfonyl fluoride in a microcentrifuge tube.
3. Incubate the immunoprecipitation tube at 4 °C for 4 h with gentle rotation. Then add 20 μg protein A agarose beads and further incubate for 1 h.
4. Centrifuge the tubes at 3,000 × *g* for 5 min at 4 °C and wash the beads with CIP buffer twice.
5. The immunoprecipitates are used for both PLDα1 activity assay and for immunoblotting to test Gα-co-immunoprecipitation. The results for PLDα1 activity assay are shown in Fig. 2.
6. For immunoblotting, dissociate the beads with 30 μl SDS-PAGE loading buffer and ~25 μg protein is resolved in 10 %

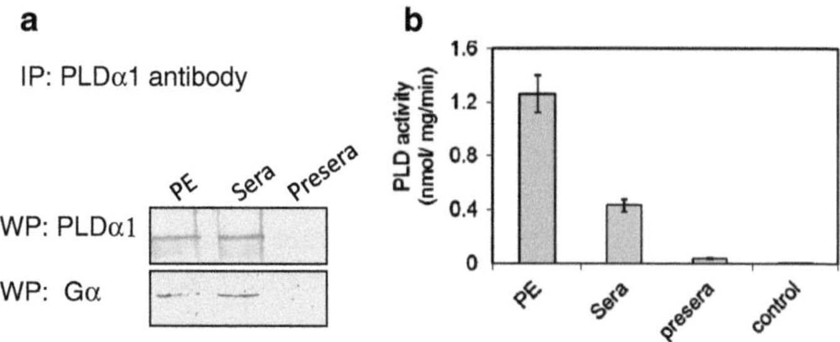

Fig. 2 Co-immunoprecipitation of Gα with PLDα1 antibody from Arabidopsis protein extract. (**a**) Co-immunoprecipitation of Gα proteins from Arabidopsis leaf protein extract by using PLDα1 antibody and protein A agarose beads. The precipitates are by resolved in SDS-PAGE and detected by immunoblotting with PLDα1 and Gα antibodies. (**b**) The precipitates are also used for measuring PLD activity

SDS-PAGE. After transferring the proteins resolved in gel to PVDF membrane, western blotting is carried out as routine way using PLDα1 and Gα antibodies.

7. Block proteins transferred from gel to PVDF membrane with 5 % nonfat milk in TBS-T buffer on a rotary shaker for 1 h. After removing blocking solution, PLDα1 and Gα antibodies are added to fresh TBS-T buffer containing 2 % nonfat milk at 1:300 ratio, respectively. Two membranes with identical samples are incubated the antibody solutions for 1 h.
8. After 15 min of washing membranes with TBS-T buffer containing 2 % for five times, the goat polyclonal secondary antibody to rabbit IgG (immunoglobulin G) conjugated with alkaline phosphatase are added to TBS-T buffer containing 2 % nonfat milk in 1:4,000 ratio, with which two membranes are incubated for 1 h.
9. Membranes are then extensively washed with TBS-T buffer for five times in 1 h. After briefly lifting to remove buffer solution, PLDα1 or Gα bands on blots are visualized by incubating the membranes with substrates from Promega. Western blotting results for Co-IP are shown in Fig. 2.

3.5 PLDα1 Activity Assay

1. Prepare the substrate solution (enough for 50 samples): add 80 μl of the 100 mg/ml non-labeled L-phosphatidylcholine and 1.25 μl ^{3}H-labeled L-3-phosphatidyl[*N*-methyl-^{3}H] choline-1,2-dipalmitoyl into a microcentrifuge tube, dry chloroform off with liquid nitrogen, and then resuspend in 1.0 ml H_2O with vortexing and sonication (*see* **Note 8**).

2. Prepare the reaction buffer (for 50 sample reactions): 100 mM MES (pH 6.0) containing 25 mM $CaCl_2$ and 0.3 mM SDS. For this, mixing 2 ml of 0.5 M MES, 0.5 ml of 0.5 M $CaCl_2$, and 0.1 ml of 30 mM SDS, and 6.4 ml H_2O.
3. Mix 160 μl reaction buffer, 20 μl enzyme sample solution, and 20 μl substrate solution in microcentrifuge tube, then vortex and incubate in 30 °C shaking water bath for 30 min.
4. Add 1 ml chloroform: methanol (2:1) and vortex vigorously, then centrifuge at 10,000 × *g* to separate the phases. Taking 150 μl out of 400 μl upper aqueous phase to scintillation liquid, and briefly vortex and count in scintillation counter.
5. PLDα1 activity can be calculated by using the specific radioactivity (0.02424 nmol/pcm) of substrates, activity (nmol/μg protein) = 0.02424 × total cpm/total protein (μg). The specific PLD activity per sample is calculated as follows: PLD activity (nmol/min/mg protein) = CPM/sample × 0.00002424 nmol/min/protein concentration.
6. Alternatively, PLDα1 activity can be assayed using a different method when the effect of Gα binding on PLD activity is measured (*see* **Note 9**). In this case, another mixed lipid vesicles composed of 3.6 μmol of phosphatidylethanolamine, 0.32 μmol of PIP_2, and 0.22 μmol of PC in the presence of 100 μM $CaCl_2$. This PC-PIP_2 method can be performed according to a procedure described elsewhere [22].

3.6 Gα GTPase Activity Assay

1. Gα GTPase activity is measured spectrophotometrically by monitoring at A_{360nm} inorganic phosphate release from GTP by GST-Gα by using the EnzChek® phosphate assay kit. Follow the manufacturer's instructions carefully and plan all activity assay reagents and timetable for assaying more samples.
2. Add the purified GST-Gα or free GST (as a control) in Tris–HCl buffer to 0.72 ml of reaction buffer (50 mM Tris–HCl, pH 7.6, and 10 mM NaCl) at various amounts (0.05–0.15 μmol) in a cuvette [10].
3. Add 0.2 ml of 2-amino-6-mercapto-7-methylpurine ribonucleotide, 10 μl (1 U) of purine nucleotide phosphorylase to the reaction mixture in the cuvette.
4. Add 10 μl of 0.2 mM GTP at a standard reaction mixture, or indicated concentrations, to bring a total volume of 1 ml.
5. Mix and start the reaction by adding 10 mM $MgCl_2$.
6. Monitor and record the absorbance changes at A_{360nm} every 5 min. The amount of phosphate released from GTP is calculated based on a standard curve that is determined with known amounts of KH_2PO_4 in the same manner.

3.7 Mutual Effects of PLDα1-Gα Interaction

1. For effects of guanine nucleotides and their analogues on PLDα1-Gα interaction, the same amounts (0, 5, 10, and 20 μM) of GDP, GTP, GDPαS, GTPαS, or Gpp(NH)p are added to the pulldown assays, either with partially purified His-PLDα1 wild-type or DRY motif mutants, or plant leaf total protein (50 μg proteins from 10,000 × *g* supernatant) extracts containing PLDα1 (*see* **Note 10**).
2. GST-Gα glutathione beads (0.05 μmol of GST-Gα) are added to parallel assays at the same concentration. Following the incubation, beads are pulled down and used for measuring PLD activity or another portion of beads is dissolved in SDS-PAGE sample buffer, and immunoblotting is conducted to detect the PLDα1 using PLDα1 antibody (*see* **Note 11**).
3. For effects of PLDα1-Gα interaction on PLDα1 activity, the partially purified His-PLDα1, wild-type or mutants (0.2 μmol), and GST-Gα at different concentrations (from 0.2 to 0.5 μmol) are co-incubated and then assay PLD activity.
4. Co-incubate the partially purified His-PLDα1, wild-type or mutants, with GST-Gα at a 1:1 molar ratio (0.18 μmol of each), then add different concentrations (5, 10, and 20 μM) of GDPαS and GTPαS. PLD activity is assayed for each of these combinations.
5. For effects of Gα-PLDα1 interaction on Gα GTPase activity, Gα GTPase activity is assayed with kit with 0.17 μmol of GST-Gα, in the presence of different concentration of His-PLDα1 or its DRY motif mutants (0.17–0.5 μmol).

3.8 Yeast Two-Hybrid Assay

1. A yeast-two hybrid system based on pEG 202 (bait plasmid) and pJG 4–5 is used for testing PLDα and Gα interaction. PLDα1 wild-type and three DRY motif mutants $PLD\alpha_{E563A}$, $PLD\alpha_{K564A}$, and $PLD\alpha_{F565A}$ cDNAs are subcloned in frame of fusion to BD (LexA) into pEG202, respectively. Gα cDNA is subcloned in frame of fusions with AD(B42) into pJG4-5 at *Eco R1* and *Xho 1* sites [11].
2. Transform competent yeast cells (*Saccharomyces cerevisiae* EGY48 containing pSH18-34) that are freshly made from overnight culture in SD-Ura broth media the sequencing-confirmed constructs by using a lithium acetate/polyethylene glycol-based method.
3. Transformation is done in two steps: First, transform yeast competent cells with pEG 202-PLDα1 wild-type and mutant constructs, and select positive clones on SD-Ura-His medium plates. The selected positive clones are confirmed by PCR and western blot for complete expression of PLDα1 wild-type and mutants. Second, transform each type of positive yeast cells

containing different bait proteins (PLDα1 wild-type and mutants) with pJG4-5-Gα, respectively, by using the same method, and select positive clones on SD-Ura-His-Trp medium plates. The selected positive clones are confirmed by RT-PCR for Gα fusion expression.

4. Test the autoactivation of co-transformed yeast host cells with different combinations of bait-prey vectors in various SC media containing glucose or galactose but missing Ura, His, Trp, or Leu, or any combination (*see* **Note 12**).
5. Test the interaction between Gα and PLDα1 and mutants with both LacZ⁻ and –Leu marker media Galactose or glucose/SC–Ura–His–Trp–Leu, or Galactose or glucose/SC–Ura–His–Trp + X-gal according to manufacturer's instruction. β-Galactosidase activity is assayed and activity is calculated and expressed through a formula,

$$1{,}000\mathrm{OD}_{420\mathrm{nm}} - 1.75\mathrm{OD}_{550\mathrm{nm}} \,/\, \text{total volume} \times \mathrm{OD}_{600\mathrm{nm}}.$$

4 Notes

1. Perhaps because phosphatidyl transferase activity of PLDα1 has deleterious effects on bacteria cells, expression level of PLDα1 in several tested types of bacteria is always low. However, its activity can be easily assayed with described methods here. To obtain enough His-taged PLDα1 proteins, two induction methods can be used in 250 ml and larger culture scales: 0.1 mM IPTG (final concentration) to bacteria cells of $\mathrm{OD}_{600\mathrm{nm}}$ 0.6–0.8 for 12 h at room temperature, or 0.4 mM IPTG (final concentration) induction of bacteria to express His-PLDα1 proteins for 6 h at 30 °C. However, for preservation of active PLDα1 proteins, lower induction temperature and longer induction time is necessary and important; induction with 0.4 mM IPTG at 16–20 °C for 16–20 h even gets better amounts of active proteins.
2. Most protein/enzyme samples in this work should be kept on ice, and experiments should be done at 4 °C, for preventing either enzyme inactivation or degradation. This is particularly important for enzyme purification, activity assay, co-immunoprecipitation, pulldown assay, and protein–protein interaction. The major reasons for failure in detecting enzyme activity, co-immunoprecipitation, or pulldown assay are enzyme inactivation and protein degradation. Although appropriate protease inhibitors can help prevent protein from degradation, they may not necessarily ensure the success in detecting protein–protein interaction. Because enzyme denaturing or inactivation may cause configuration/conformation changes, which affect their interaction with other partners.

3. Our constructed GST-Gα fusion can be very easily induced to large amounts. Optimal induction of GST-Gα fusion is 0.2 mM IPTG (final concentration) induction at room temperature for 16 h.
4. PMSF should be freshly prepared in methanol. PMSF easily and completely degrades at room temperature, and can slowly degrade on ice. Otherwise, storage of PMSF in methanol at −20 °C is best. It is very effective to inhibit most major proteases in bacteria and plant extracts.
5. The primary polyclonal PLDα1 antibody is excellent for Western blotting, immunoprecipitation, and immunofluorescence. Similarly, the primary polyclonal Gα1 antibody is also excellent for both Western blotting and immunoprecipitation experiments. Their specificity has been well addressed [22, 24].
6. The primary antibodies in TBS-T containing 2 % nonfat milk can be kept safe in 4 °C for several weeks by adding 0.02 % (w/v) sodium azide (final concentration, can be made from 10 % stock sodium azide solution) for subsequent experiments.
7. Although other Arabidopsis PLD isoforms such as PLDα1 show higher similarity on DRY motif with mammalian GCPRs, they may not interact in the same way with PLDα1, since Arabidopsis sole canonical Gα also has its own ways to function in plant with its partners, which are different from its mammalian counterparts. Furthermore, this major and dominant PLD isoform, PLDα1 shares similar gene expression patterns and biological functions with Gα, not only in stomata movement [11, 25], but also in leaf water loss [25, 26], and seed germination processes (ref. 20; unpublished data).
8. Because that the presence of SDS and high levels of Ca^{2+} that might interfere with the PLD-Gα interaction, we used a different method to test how Gα-PLDα1 interaction affects PLDα1 activity. Actually, it can be observed that SDS and $CaCl_2$ indeed negatively affect their interaction.
9. PLDα1 is the major plant PLD isoform. Its expression and enzymatic activity are high and strong in universal tissues from root to pollen, petiole to petal. Its featured activity, phosphatidyl transferase activity, is involved in many physiological processes. Its activity can be measured in different means and under different conditions. In particular, measuring phosphatidyl transferase activity of PLDα1 in the presence of 25 mM $CaCl_2$ and 0.3 mM SDS, gives a strong, clear, distinguishable, and reproducible results [22].
10. The presence of GTPαS decreases Gα-PLDα1 interaction and releases Gα inhibition of PLDα1 activity; whereas GDPαS

promotes Gα-PLDα1 interaction but thereby inhibit. Therefore it enhances Gα inhibition on PLDα1 activity. This could explain the often observed phenomena that G-protein activators promote PLDα1 activity in many assays. PLDα1 binding promotes Gα GTPase activity, and intend to form the Gα-GDP-PLDα1 complex (GDP-bound form of Gα to release Gβγ). Therefore, PLDα1 seems to have GTPase-activating protein activity to promote the exchange of GDP for GTP, which is interesting and worthy to be further tested.

11. Excellent antibodies for PLDα1 and Gα are also important prerequisites for studying their interaction. Specificity of these antibodies contributes largely to the accuracy of interaction detection. In addition, both PLDα1 and Gα have strong and unique activities that can be easily measured. This facilitates the functional study of their interaction and biological significance.
12. pJG4-5-Gα and pEG202-PLDα1 wild-type and mutant constructs in yeast cells EGY48 appear do not show any auto-activation in both yeast growth in various SC media containing glucose or galactose, and limited amino acids.

Acknowledgments

The work was supported by the Fundamental Research Funds for the Central Universities (Project 2013PY065) to JZ and the National Science Foundation (IOS-0818740) to XW. We thank Brian Fanella for critically reading the manuscript.

References

1. Wang X, Xu L, Zheng L (1994) Cloning and expression of phosphatidylcholine-hydrolyzing phospholipase D from *Ricinus communis* L. J Biol Chem 269:20312–20317
2. Wang X (2000) Multiple forms of phospholipase D in plants: the gene family, catalytic and regulatory properties, and cellular functions. Prog Lipid Res 39:109–149
3. Wang X (2005) Regulatory functions of phospholipase D and phosphatidic acid in plant growth, development, and stress responses. Plant Physiol 139:566–573
4. Li M, Hong Y, Wang X (2009) Phospholipase D- and phosphatidic acid-mediated signaling in plants. Biochim Biophys Acta 179:927–935
5. Munnik T, Arisz SA, De Vrije T, Musgrave A (1995) G Protein activation stimulates phospholipase D signaling in plants. Plant Cell 7:2197–2210
6. Rovati GE, Capra V, Neubig RR (2007) The highly conserved DRY motif of class A G protein-coupled receptors: beyond the ground state. Mol Pharmacol 71:959–964
7. Lee CS, Kim S, Park JB, Lee MN, Lee HY, Suh P-G, Ryu SH (2006) The phox homology domain of phospholipase D activates dynamin GTPase activity and accelerates EGFR endocytosis. Nat Cell Biol 8:477–484
8. Henage LG, Exton JH, Brown HA (2006) Kinetic analysis of a mammalian phospholipase D: allosteric modulation by monomeric GTPases, protein kinase C, and polyphosphoinositides. J Biol Chem 281:3408–3417
9. Exton JH (1999) Regulation of phospholipase D. Biochim Biophys Acta 1439:121–133

10. Zhao J, Wang X (2004) Arabidopsis phospholipase Dα1 interacts with the heterotrimeric G-protein α-subunit through a motif analogous to the DRY motif in G-protein-coupled receptors. J Biol Chem 279:1794–1800
11. Mishra G, Zhang W, Deng F, Zhao J, Wang X (2006) A bifurcating pathway directs abscisic acid effects on stomatal closure and opening in Arabidopsis. Science 312:264–266
12. Peng HJ, Henkels KM, Mahankali M, Dinauer MC, Gomez-Cambronero J (2011) Evidence for two CRIB domains in phospholipase D2 (PLD2) that the enzyme uses to specifically bind to the small GTPase Rac2. J Biol Chem 286:16308–16320
13. Adjobo-Hermans MJ, Goedhart J, Gadella TW Jr (2006) Plant G protein heterotrimers require dual lipidation motifs of Gα and Gγ and do not dissociate upon activation. J Cell Sci 119:5087–5097
14. Wang S, Assmann SM, Fedoroff NV (2008) Characterization of the Arabidopsis heterotrimeric G protein. J Biol Chem 283:13913–13922
15. Jones JC, Temple BR, Jones AM, Dohlman HG (2011) Functional reconstitution of an atypical G protein heterotrimer and regulator of G protein signaling protein (RGS1) from *Arabidopsis thaliana*. J Biol Chem 286:13143–13150
16. Pandey S, Assmann SM (2004) The Arabidopsis putative G protein-coupled receptor GCR1 interacts with the G protein alpha subunit GPA1 and regulates abscisic acid signaling. Plant Cell 16:1616–1632
17. Pandey S, Nelson DC, Assmann SM (2009) Two novel GPCR-type G proteins are abscisic acid receptors in Arabidopsis. Cell 136:136–148
18. Chen JG, Willard FS, Huang J, Liang J, Chasse SA, Jones AM, Siderovski DP (2003) A seven-transmembrane RGS protein that modulates plant cell proliferation. Science 301:1728–1731
19. Huang J, Taylor JP, Chen JG, Uhrig JF, Schnell DJ, Nakagawa T, Korth KL, Jones AM (2006) The plastid protein THYLAKOID FORMATION1 and the plasma membrane G-protein GPA1 interact in a novel sugar-signaling mechanism in Arabidopsis. Plant Cell 18:1226–1238
20. Lapik Y, Kaufman LS (2003) The Arabidopsis cupin domain protein AtPirin1 interacts with the G protein alpha-subunit GPA1 and regulates seed germination and early seedling development. Plant Cell 15:1578–1590
21. Warpeha KM, Lateef SS, Lapik Y, Anderson M, Lee BS, Kaufman LS (2006) G-protein-coupled receptor 1, G-protein Gα-subunit 1, and prephenate dehydratase 1 are required for blue light-induced production of phenylalanine in etiolated Arabidopsis. Plant Physiol 140:844–855
22. Pappan K, Wang X (1999) Plant phospholipase Dα is an acidic phospholipase active at near-physiological Ca^{2+} concentrations. Arch Biochem Biophys 368:347–353
23. Weiss CA, Huang H, Ma H (1993) Immunolocalization of the G protein α. subunit encoded by the GPA1 gene in Arabidopsis. Plant Cell 5:1513–1528
24. Ma H, Yanofsky MF, Meyerowitz EM (1990) Molecular cloning and characterization of GPA1, a G protein a subunit gene from *Arabidopsis thaliana*. Proc Natl Acad Sci USA 87:3821–3825
25. Wang XQ, Ullah H, Jones AM, Assmann SM (2001) G protein regulation of ion channels and abscisic acid signaling in Arabidopsis guard cells. Science 292:2070–2072
26. Sang Y, Zheng S, Li W, Huang B, Wang X (2001) Regulation of plant water loss by manipulating the expression of phospholipase Dα. Plant J 28:135–144

Chapter 4

Analysis of Cell Division and Cell Elongation in the Hypocotyls of Arabidopsis Heterotrimeric G Protein Mutants

Zhaoqing Jin, Wellington Muchero, and Jin-Gui Chen

Abstract

The heterotrimeric G-proteins form classical signal transduction complexes conserved in all eukaryotes. The repertoire of G-protein signaling complex is much simpler in plants than in metazoans. One of the best understood functions for the plant G-protein complex is its modulation of cell proliferation. A method is described here to quantify cell division and cell elongation in the Arabidopsis heterotrimeric G-protein mutants using the hypocotyl as a model system.

Key words Arabidopsis, Cell division, Cell elongation, Heterotrimeric G-proteins, Hypocotyl

1 Introduction

Signaling through the heterotrimeric guanine nucleotide-binding proteins (G-proteins) is a conserved mechanism found in fungi, animals and plants. Heterotrimeric G-proteins consist of three subunits, namely, G-protein α (Gα), β (Gβ), and γ (Gγ) subunits. G-proteins couple the recognition of extracellular signals by cell surface seven-transmembrane G-protein-coupled receptors (GPCRs) to the activation of downstream effectors [1]. Upon ligand binding, GPCR activates G-protein signaling by promoting the exchange of Gα-bound GDP for GTP and subsequently the dissociation of the Gβγ dimer from the Gα. The activated Gα (GTP bound) and the Gβγ dimer (freely released) then activate downstream effector proteins. The Gα subunit has intrinsic GTPase activity and will eventually hydrolyze the attached GTP to GDP, allowing its re-association with the Gβγ dimer. The intrinsic GTPase activity of the Gα subunit can be accelerated by the Regulator of G-protein Signaling (RGS) proteins which negatively regulate G-protein signaling [2].

Mark P. Running (ed.), *G Protein-Coupled Receptor Signaling in Plants: Methods and Protocols*, Methods in Molecular Biology, vol. 1043, DOI 10.1007/978-1-62703-532-3_4, © Springer Science+Business Media, LLC 2013

The repertoire of G-protein signaling complex is much simpler in plants than in metazoans. The model plant species, *Arabidopsis thaliana*, contains only one canonical Gα (GPA1), one Gβ (AGB1), three Gγ (AGG1, AGG2 and AGG3) subunits, and one RGS protein (AtRGS1), and no bona fide seven-transmembrane GPCR together with its ligand has been unequivocally identified [3–7]. The smaller repertoire of the G-protein signaling complex in plants offers a unique advantage over its counterpart in mammals for dissecting their functions. Molecular and genetic analysis demonstrated that G-proteins play regulatory roles in diverse developmental processes in plants [4, 6]. One of the best understood functions for the G protein complex is its modulation of cell proliferation [8]. For example, the etiolated *gpa1* mutant seedlings have short hypocotyls due to reduced epidermal cell division, and the light-grown *gpa1* seedlings produce fewer lateral roots due to reduced lateral root primordia formation [9, 10]. Similar to *gpa1*, the etiolated *agb1* mutant seedlings also have short hypocotyls due to reduced axial epidermal cell division [10, 11]. However, opposite to *gpa1*, light-grown *agb1* seedlings produce more lateral roots largely due to increased lateral root primordia formation [10, 11].

The Arabidopsis hypocotyl is a model system to study cell division and cell elongation, because the number of epidermal cells in a single cell file from the base to the top of a hypocotyl is determined during embryogenesis. In dark- or light-grown Arabidopsis seedlings, cell divisions are absent or insignificant in the epidermal or cortical cells of the elongating hypocotyls [12]. Therefore, by counting the number (indicator of cell division) and measuring the length (indicator of cell elongation) of epidermal or cortical cells in a single cell file longitudinally from the base to the top of a hypocotyl, one can study cell division (occurred during embryogenesis) and cell elongation (occurred post-embryogenesis) simultaneously. In this report, we provide a detailed method for analyzing cell division and cell elongation in Arabidopsis hypocotyls. We use G-protein mutants as examples but this method can also be applied to any other Arabidopsis mutants. For example, by using this method, it has recently been determined that Brassinosteroid (BR) biosynthesis mutants, *det2-1* and *dwf4-102*, and signaling mutants, *bri1-5* and *bri1-4*, have defects in both cell elongation and cell division [13].

2 Materials

2.1 Arabidopsis Seeds

Wild-type: Columbia (Col) and Wassilewskia (Ws); G-protein mutants: *gpa1-1* and *gpa1-2* (in Ws) [9], *gpa1-3* and *gpa1-4* (in Col) [14], *agb1-1* (in Col) [15], *agb1-2* (in Col) [11], *agg1-1* (in Ws) and *agg2-1* (in Col) [16], and *Atrgs1-1* and *Atrgs1-2* (in Col) [17].

2.2 Sterilizing Reagents

80 % (v/v) ethanol; 35 % (v/v) bleach with 0.05 % (v/v) tween-20; and sterile distilled H_2O.

2.3 MS Medium

Murashige and Skoog (MS) basal medium with vitamins (PlantMedia, Dublin, OH, USA), containing 1 % (w/v) sucrose, 0.5 % (w/v) phytoagar (PlantMedia), and pH adjusted to 7.5 with 1 N KOH. Petri dishes. Aluminum foil.

2.4 Growth Chamber

Conditions set at approximately 125 μmol/m^2/s with a 14/10 h photoperiod, 22–24 °C.

2.5 Clearing Solution

Chloral hydrate solution (chloral hydrate–glycerol–water = 8:1:3) (w/v/v).

2.6 Microscope

Compound microscope with Differential Interference Contrast (DIC) equipped with a digital image acquisition and processing system.

3 Methods

3.1 Seed Germination

The quality of seeds impacts germination and subsequently the growth of hypocotyl. Therefore, seed germination quality control is necessary to obtain proper plant materials for hypocotyl cell division and cell elongation assays. Wild-type and mutant seeds should be harvested from plants grown under identical conditions and stored identically. Seeds of comparable, matched lots are sterilized with 80 % ethanol for 2 min, followed by 35 % bleach plus tween-20 for 5 min, and washed five times with sterile distilled H_2O. Sterilized seeds are then sown individually (*see* **Note 1**) on the petri dish plates containing sterilized MS medium (*see* **Note 2**) and subjected to stratification treatment at 4 °C in darkness for 72 h (*see* **Note 3**). Then, the imbibed seeds are exposed to light (~125 μmol/m^2/s) for 12 h and the plates are wrapped with two layers of aluminum foil and moved into a growth chamber (*see* **Note 4**) at 22–24 °C for 60 h (*see* **Note 5**).

3.2 Sample Clearing

After seeds have been cultured at 22–24 °C for 60 h (*see* **Note 6**), MS plates are opened under lab conditions. Etiolated seedlings are picked up carefully from MS medium and placed immediately into chloral hydrate solution (*see* **Note 7**) in 1.5-ml eppendorf microtubes (*see* **Note 8**). Pay attention not to damage hypocotyls by holding gently the cotyledons of etiolated seedling using forceps. It is also necessary to wear gloves when handling chloral hydrate solution. The seedlings should be cleared in the chloral hydrate solution for at least 48 h, and can be stored in the solution for several months.

3.3 Microscopy

After being cleared with chloral hydrate solution, samples are ready to be transferred onto slides for microscopic observation (*see* **Note 9**). Because chloral hydrate-cleared etiolated seedlings often appear transparent and fragile, extra care should be taken when transferring seedlings to slides. Put one drop of chloral hydrate solution onto the center of each slide prior to seedling transferring. Avoid generating bubbles when placing cover slide (*see* **Note 10**). First use 5× or 10× objective to locate the hypocotyl–root junction (*see* **Note 11**) and the shoot apical meristem (*see* **Note 12**). Then, use a 20× objective to observe epidermal cells starting from the hypocotyl–root junction.

3.4 Counting Cell Number and Measuring Cell Length

Epidermal cells are often protruding and have oblique ends in a longitudinal direction, whereas cortex cells often have blunt ends (*see* **Note 13**). Because epidermal cells occupy the outmost layer of the hypocotyl, they can be easily located by adjusting the focal plane carefully. Because some epidermal cells may display twisted growth, it may be necessary to adjust the focal plane frequently. It is important to trace a single cell file longitudinally from the hypocotyl–root junction to the top of hypocotyl, a position that is lateral to cotyledons and ends at the site of the shoot apical meristem. In order to measure cell length, it is necessary to take snapshot of each image, paying attention to the boundary between images. Use the measurement tool of the digital image acquisition and processing system (e.g., we used a Leica DM-6000B upright microscope with phase and DIC equipped with a Leica FW4000 digital image acquisition and processing system [Leica Microsystems (Canada) Inc., Richmond Hill, ON, Canada]) to record the length from bottom to top of each epidermal cell sequentially from the hypocotyl–root junction to the top of a hypocotyl. The first cell is the one next to hypocotyl–root junction and the last cell is the one next to the shoot apical meristem (lateral to cotyledons). Typically, a single epidermal cell file longitudinally from the base to the top of a hypocotyl consists of 19–21 cells in a 2.5-day-old wild-type Arabidopsis etiolated seedling (*see* **Note 14**).

3.5 Data Mining

We typically count cell number and measure cell length in ten seedlings for each genotype. If all seedlings for the same genotype contain the same number of epidermal cells in a single cell file from the base to the top of a hypocotyl, statistical analysis can be performed to calculate the average cell length in each position with standard error (SE) (Fig. 1). In some cases where seedlings in a given genotype do not contain the exact same number of epidermal cells (e.g., 19 or 21), it may be necessary to plot the length of each epidermal cell (*y*-axis) against its position (*x*-axis) in a hypocotyl individually. In 2.5-day-old wild-type etiolated seedlings, the longest epidermal cell is generally located in the middle whereas the shortest epidermal cell is located on the base or top of a

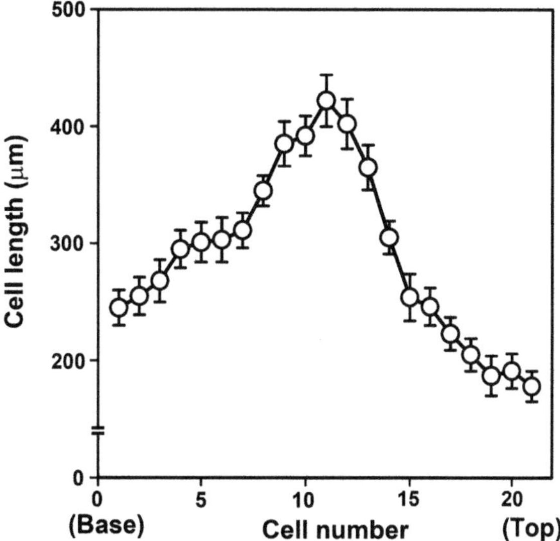

Fig. 1 The number and length of hypocotyl epidermal cells in 2.5-day-old wild-type (Col) Arabidopsis etiolated seedlings. Shown are the average lengths of epidermal cells in a single cell file longitudinally from the base to the top of hypocotyls from ten etiolated seedlings ± SE

hypocotyl. The plot of cell length vs. position displays a bell-shaped profile (Fig. 1). In some mutants, particularly in dwarf mutants, however, such bell-shaped profiles are not observed. For example, a BR-deficient mutant, *det2-1*, displayed a flat profile with most epidermal cells having similar length [13].

4 Notes

1. It is important to sow seeds individually on MS medium to avoid contact between seedlings that affects hypocotyl growth dynamics and orientation.
2. Because seed germination can vary from plate to plate, it is necessary to germinate and grow all genotypes in the same MS medium plate (if possible) to avoid variation.
3. Seeds can be subjected to stratification treatment at 4 °C in darkness for 2–5 days.
4. It is always easier to sample and examine seedlings with straight hypocotyls, so it is necessary to ensure that MS plates are placed on a flat surface.
5. For most genotypes, 2.5-day (60 h)-old etiolated seedlings are suitable. Although >4-day-old etiolated seedling can also be used, it is often difficult to measure cell length because the

epidermal cells can reach more than 800 μm in length which requires taking several image snapshots (under 20× objective) for one single epidermal cell.

6. The age of etiolated seedlings does not include the 12 h light treatment.
7. Chloral hydrate is a controlled substance in the United States of America. A certificate or prescription may be required to purchase this chemical.
8. Because chloral hydrate solution can dehydrate very quickly, it is important to tightly close its container and sampling eppendorf microtubes.
9. It is essential to use DIC for locating, counting and measuring hypocotyl cells.
10. Because of the effects of chloral hydrate dehydration, it is often difficult to mount the sample slides. One may consider transferring the samples back to eppendorf microtubes for storage.
11. The base of a hypocotyl is next to the hypocotyl–root junction which can be determined by the appearance of root hair cells.
12. The top of a hypocotyl is next to the apical meristem and is also lateral to cotyledons. The petiole of cotyledons grows in a direction that forms an angle with the hypocotyl axis. It may be necessary to adjust the focal plane carefully around this region to locate the apical meristem and the cotyledon petiole.
13. Because epidermal cells on the plant surface have a more important role in determining the shape of an organ than other cell types and epidermal cells on hypocotyls can be easily recognized due to their location and shape, this method is described for observing epidermal cells on hypocotyls but it can also be readily adjusted to observe other cell types, such as cortex cells. Arabidopsis hypocotyls have two layers of cortex cells, outer and inner cortex cells. The outer cortex cells are located just beneath the epidermal cells and are relatively easier to observe by adjusting the focal plane carefully.
14. This method is described for counting cell number and measuring cell length in etiolated Arabidopsis seedlings but it can also be readily used for light-grown seedlings. Because light inhibits hypocotyl elongation, resulting in shorter hypocotyls, it may be necessary to apply dim-light conditions (e.g., 30 μmol/m^2/s), particularly for dwarf mutants that display extreme short hypocotyl phenotypes under normal light conditions. For example, BR-deficient mutant, *det2-1*, and signaling mutant, *bri1-4*, have very short hypocotyls under normal light conditions (e.g., 125 μmol/m^2/s). When they are grown under dim-light conditions, their hypocotyls are much longer, making it easier and more accurate to quantify the number and length of epidermal cells [13].

Acknowledgments

This work was supported by the Laboratory Directed Research and Development Program of Oak Ridge National Laboratory. Oak Ridge National Laboratory is managed by UT-Battelle, LLC, for the US Department of Energy under Contract No. DE-AC05-00OR22725.

References

1. Gilman AG (1987) G proteins: transducers of receptor-generated signals. Annu Rev Biochem 56:615–649
2. Neubig RR, Siderovski DP (2002) Regulators of G-protein signalling as new central nervous system drug targets. Nat Rev Drug Discov 1:187–197
3. Assmann SM (2002) Heterotrimeric and unconventional GTP binding proteins in plant cell signaling. Plant Cell 14(Suppl):S355–S373
4. Perfus-Barbeoch L, Jones AM, Assmann SM (2004) Plant heterotrimeric G protein function: insights from Arabidopsis and rice mutants. Curr Opin Plant Biol 7:719–731
5. Temple BRS, Jones AM (2007) The plant heterotrimeric G-protein complex. Annu Rev Plant Biol 58:249–266
6. Chen JG (2008) Heterotrimeric G proteins in plant development. Front Biosci 13: 3321–3333
7. Urano D, Jones JC, Wang H, Matthews M, Bradford W, Bennetzen JL, Jones AM (2012) G protein activation without a GEF in the plant kingdom. PLoS Genet 8(6):e1002756
8. Chen JG (2010) Heterotrimeric G-proteins and cell division in plants. In: Yalovsky S, Baluska F, Jones AM (eds) Integrated G proteins signaling in plants. Springer, Heidelberg, pp 155–176
9. Ullah H, Chen JG, Young JC, Im KH, Sussman MR, Jones AM (2001) Modulation of cell proliferation by heterotrimeric G protein in Arabidopsis. Science 292:2066–2069
10. Chen JG, Gao Y, Jones AM (2006) Differential roles of Arabidopsis heterotrimeric G-protein subunits in modulating cell division in roots. Plant Physiol 141:887–897
11. Ullah H, Chen JG, Temple B, Boyes DC, Alonso JM, Davis KR, Ecker JR, Jones AM (2003) The β-subunit of the Arabidopsis G protein negatively regulates auxin-induced cell division and affects multiple developmental processes. Plant Cell 15:393–409
12. Gendreau E, Traas J, Desnos T, Grandjean O, Caboche M, Hofte H (1997) Cellular basis of hypocotyl growth in *Arabidopsis thaliana*. Plant Physiol 114:295–305
13. Gao Y, Wang S, Asami T, Chen JG (2008) Loss-of-function mutations in the Arabidopsis heterotrimeric G-protein α subunit enhance the developmental defects of brassinosteroid signaling and biosynthesis mutants. Plant Cell Physiol 49:1013–1024
14. Jones AM, Ecker JR, Chen JG (2003) A reevaluation of the role of the heterotrimeric G protein in coupling light responses in Arabidopsis. Plant Physiol 131:1623–1627
15. Lease KA, Wen J, Li J, Doke JT, Liscum E, Walker JC (2001) A mutant Arabidopsis heterotrimeric G-protein β subunit affects leaf, flower, and fruit development. Plant Cell 13:2631–2641
16. Trusov Y, Rookes JE, Tilbrook K, Chakravorty D, Mason MG, Anderson D, Chen JG, Jones AM, Botella JR (2007) Heterotrimeric G protein γ subunits provide functional selectivity in Gβγ dimer signaling in Arabidopsis. Plant Cell 19:1235–1250
17. Chen JG, Willard FS, Huang J, Liang J, Chasse SA, Jones AM, Siderovski DP (2003) A seven-transmembrane RGS protein that modulates plant cell proliferation. Science 301: 1728–1731

Chapter 5

Aequorin Luminescence-Based Functional Calcium Assay for Heterotrimeric G-Proteins in Arabidopsis

Kiwamu Tanaka, Jeongmin Choi, and Gary Stacey

Abstract

Heterotrimeric GTP-binding proteins (G-proteins) and G-protein-coupled receptors are important signaling components in eukaryotes. In plants, the G-proteins are involved in diverse physiological processes, some of which are exerted via changes in the level of cytosolic free calcium concentration ($[Ca^{2+}]_{cyt}$). Various techniques have been developed to measure the change of $[Ca^{2+}]_{cyt}$, e.g., calcium-sensitive microelectrodes, chemical fluorescent dyes, and biosensors based on luminescent or fluorescent indicators. In this chapter, we describe a protocol for in vivo $[Ca^{2+}]_{cyt}$ measurement in G-protein mutants expressing aequorin, a luminescent-based calcium biosensor, to extend our knowledge about G-protein mediated Ca^{2+} signaling. This method is also applicable to other early signaling events that are mediated by changes in $[Ca^{2+}]_{cyt}$ levels.

Key words Aequorin, Coelenterazine, Cytosolic free calcium ion, Heterotrimeric GTP-binding proteins, Arabidopsis

1 Introduction

Heterotrimeric GTP binding proteins (G-proteins) are well-characterized molecular switches that are activated in response to various extracellular stimuli [1, 2]. Generally, upon signal reception by G-protein-coupled receptors, the Gα subunit dissociates from the Gβγ dimer, and either one or both of these components interact with downstream elements to transmit the signal [3]. Like animals, plants use signal transduction pathways based on G-proteins to regulate many aspects of plant growth and development occasionally via Ca^{2+} signaling. Pharmacological studies targeting the Gα subunit or genetic studies using mutants of the Gα subunit and Gβ subunit have revealed that the G-proteins participate in seed germination, pollen germination, stomatal closure, microbe elicitor responses, etc. by modulation of cytosolic free calcium concentration ($[Ca^{2+}]_{cyt}$) [4–7]. This suggests that the

Mark P. Running (ed.), *G Protein-Coupled Receptor Signaling in Plants: Methods and Protocols*, Methods in Molecular Biology, vol. 1043, DOI 10.1007/978-1-62703-532-3_5, © Springer Science+Business Media, LLC 2013

dynamics of $[Ca^{2+}]_{cyt}$ are closely linked to G-protein activity and can be excellent cellular markers to measure the G-protein-mediated signaling in plants.

There are many approaches to measure $[Ca^{2+}]_{cyt}$ such as calcium-sensitive microelectrodes, chemical fluorescent dyes, and biosensors based on luminescent or fluorescent indicators. Luminescent techniques using photoproteins (e.g., aequorin, etc.) have been extensively used to measure $[Ca^{2+}]_{cyt}$ dynamics in plants. Although an extremely sensitive light detector, such as photomultiplier tube or photon-counting camera, is required due to very low light yield, luminescent calcium reporters have several advantages over fluorescent dyes. For example, photo damage associated with excitation illumination is avoided, since the luminescent light emission is not dependent on optical excitation. Luminescence also has an intrinsically high signal-to-background ratio, because there is no autofluorescence and relatively little endogenous, background luminescence. The ability to transgenically express aequorin in plants has made this reporter particularly attractive to plant researchers to measure the cellular Ca^{2+} response to a variety of external stimuli.

Aequorin is a 21.4 kDa photoprotein originally isolated from the coelenterate jellyfish *Aequorea victoria* [8]. In order to be luminescent, the enzyme requires a hydrophobic prosthetic group, coelenterazine, to convert the apo-aequorin to aequorin, which is the active form of the enzyme. Upon Ca^{2+} binding, aequorin oxidizes coelenterazine into coelenteramide with the production of CO_2 and emission of blue light with a wavelength of approximately 470 nm (Fig. 1). The affinity of aequorin for Ca^{2+} is in the low micromolar range, and the enzymatic activity is proportional to cellular Ca^{2+} concentration in the physiological range of 50 nM to 50 μM [9]. Therefore, measurement of light emitted upon oxidation of coelenterazine by aequorin provides reliable quantification of cellular calcium levels. These measurements are comparable to that determined with fluorescent dyes [9].

In this chapter, we describe an in vivo $[Ca^{2+}]_{cyt}$ measurement technique using transgenic Arabidopsis expressing aequorin [10]. By way of example, we compare the calcium response of Arabidopsis G-protein mutants to those of wild-type plants [11]. Since the Gα subunit (GPA1) and Gβ subunit (AGB1) are encoded by single-copy genes in the Arabidopsis genome, analysis of single mutants in these genes provides a rigorous means to study their involvement in various signaling pathways. In order to trigger $[Ca^{2+}]_{cyt}$ changes, plants were treated with either extracellular ATP [12] or other microbial elicitors (Figs. 2 and 3). The method described here is generic and applicable for other studies using different mutant backgrounds that exhibit a change in the $[Ca^{2+}]_{cyt}$ levels [13–18].

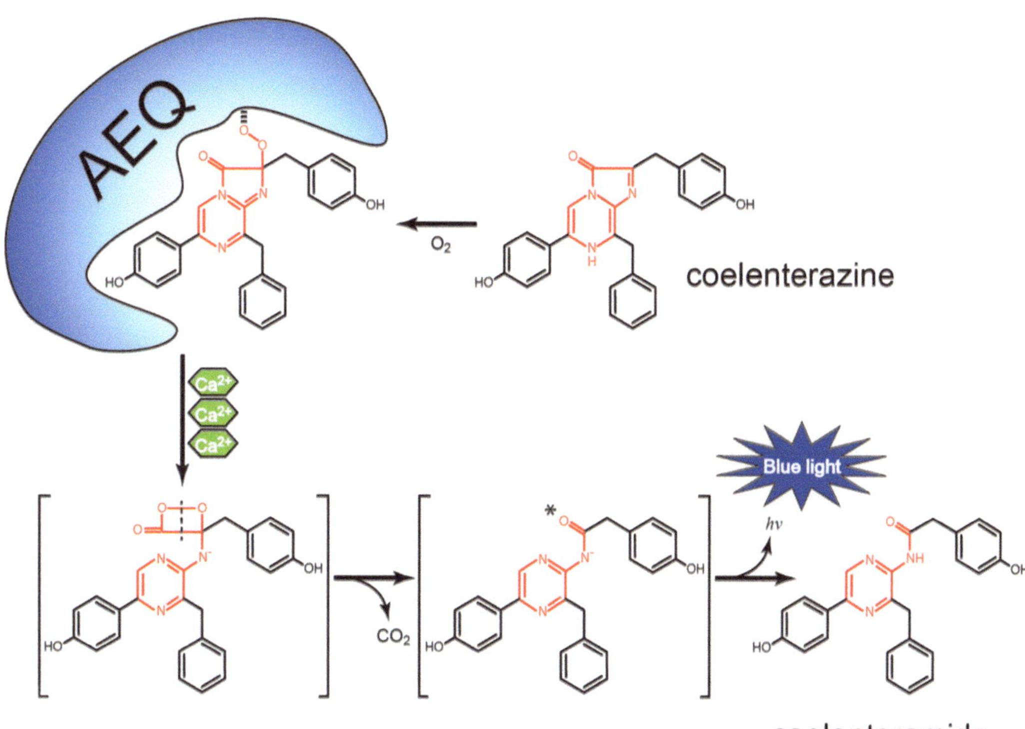

Fig. 1 Diagram showing Ca^{2+}-dependent light generation by aequorin. Apo-aequorin is converted to the active form, aequorin (AEQ), when reconstituted by a luminophore coelenterazine in the presence of oxygen (O_2). Upon binding of three molecules of Ca^{2+} to the respective EF-hands on the aequorin protein, coelenterazine is oxidized and cyclized to give the dioxetanone intermediate, followed by a conformational change of the protein accompanied by the release of carbon dioxide (CO_2) while producing the singlet-excited coelenteramide (*asterisk*) that emits *blue* light ($hv_{\sim 470nm}$). The part of the coelenterazine molecule where the changes occur is indicated with *red* color

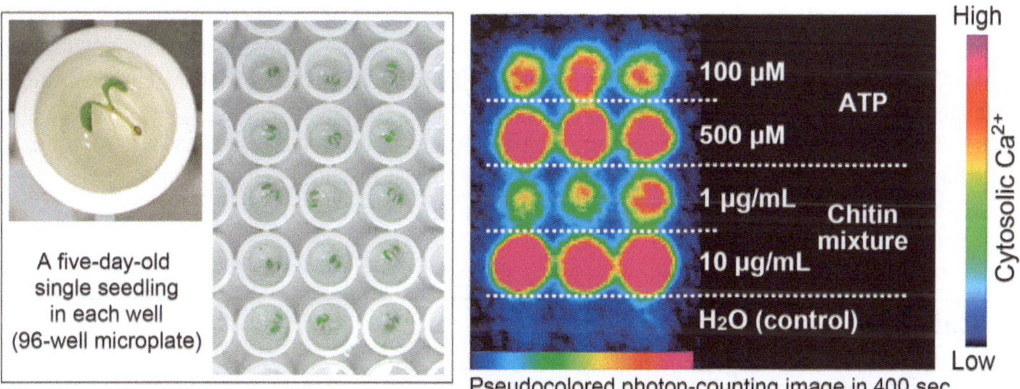

Fig. 2 An example of aequorin-based Ca^{2+} measurement in a 96-well white plate. *Left* picture shows the individual aequorin Arabidopsis seedlings in each well of the microplate. The *right* picture shows a pseudocolored photon-counting image (integrated over 400 s) after addition of ATP or chitin mixture. $[Ca^{2+}]_{cyt}$ has been coded according to the scale

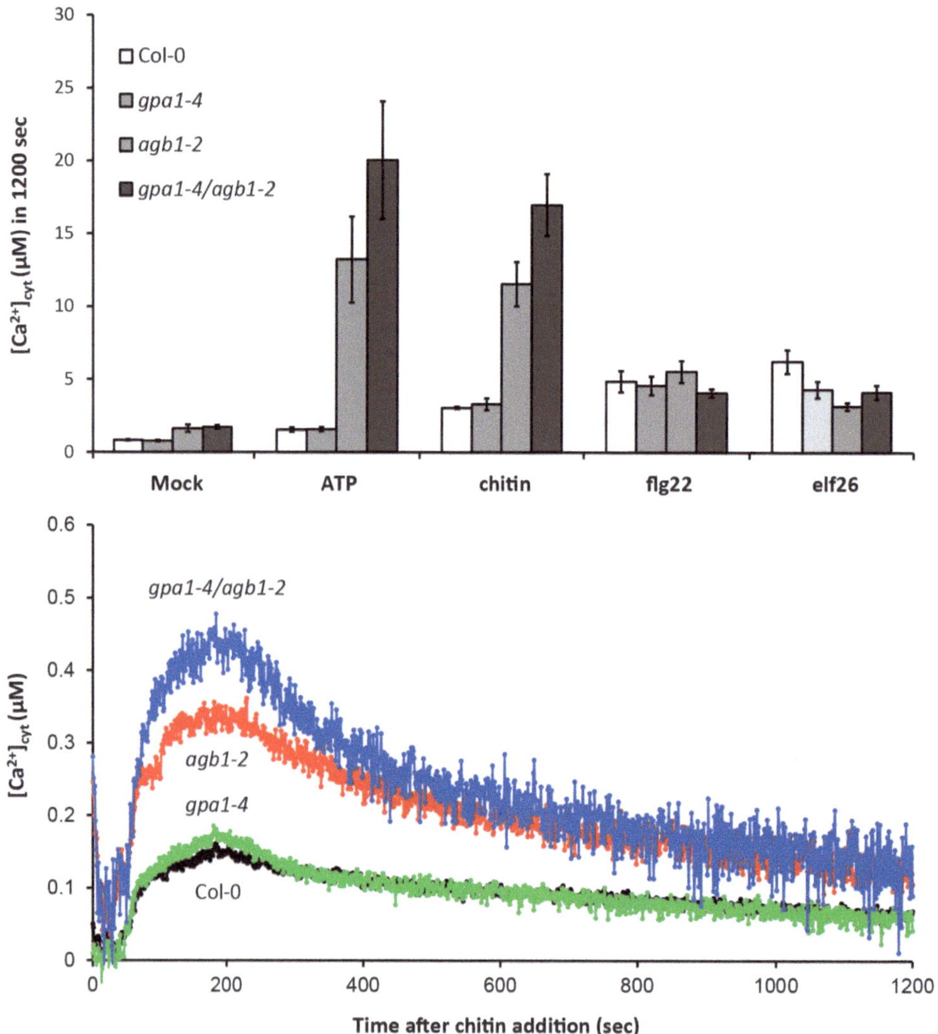

Fig. 3 Measurement of stimuli-induced $[Ca^{2+}]_{cyt}$ changes using Arabidopsis in G-protein mutants expressing aequorin. Histogram represents integrated $[Ca^{2+}]_{cyt}$ values over 1,200 s after various stimuli treatments: mock (2 mM MES buffer), ATP 100 μM, chitin mixture 10 μg/mL, flg22 100 nM, and elf26 100 nM. Each value represents the mean of six seedlings ±SE. *Line graph* shows kinetic differences in $[Ca^{2+}]_{cyt}$ responses to chitin treatment. Note that the *agb1* mutant seedlings exhibited hyper-responsiveness to extracellular ATP or chitin, suggesting that the Gβ-subunit is negatively involved in the cellular response to these stimuli

2 Materials

2.1 Plant Materials and Plant Growth

1. Transgenic Arabidopsis expressing the photoprotein apo-aequorin (*see* **Note 1**).
2. Heterotrimeric G-protein mutants confirmed by PCR-based genotyping using gene specific primers (*see* **Note 2**).

3. Half-strength Murashige and Skoog (MS) medium containing 1 % (w/v) sucrose, 1 % (w/v) agar, 0.05 % (w/v) MES (pH 5.7), and Gamborg's B5 vitamins.
4. Growth chamber set at 22 °C under long-day conditions (16 h light at ~80 μE/m^2/s and 8 h dark).

2.2 Aequorin Luminescence-Based Calcium Assay

1. Coelenterazine stock solution: 1 mM native coelenterazine (Nanolight Technology) dissolved in ethanol. This is 100× stock solution and diluted to the specific concentration needed. This solution should be stored in the dark at −20 °C and, under these conditions, is stable up to two weeks (*see* **Note 3**).
2. Reconstitution buffer: 2 mM MES (pH 5.7 with 10 N KOH), 10 mM $CaCl_2$. Immediately prior to use, add coelenterazine from the stock solution (**item 1** of Subheading 2.2) to a final concentration of 10 μM (*see* **Note 4**).
3. 96-well white polystyrene microplates with opaque flat bottom (*see* **Note 5**).
4. A CCD camera with a fully light-tight dark box (Photek 216; Photek, Ltd.) (*see* **Note 6**).
5. Multichannel pipettes: to apply the chemical solutions to multiple wells at the same time.
6. Discharging solution: 2 M $CaCl_2$ dissolved in 20 % (v/v) ethanol.

3 Methods

3.1 Preparation of the Aequorin Transgenic Lines in the G-Protein Mutant Background Through a Genetic Cross

To transfer an aequorin gene into the G-protein mutants, F_2 generation pools from a cross between each mutant and the aequorin transgenic Arabidopsis line are subjected to screening as follows. T-DNA insertion of the mutants is evaluated by PCR-based genotyping. Introgression of the aequorin gene is confirmed by bioluminescence of aequorin itself (*see* **Note 7**).

1. Grow the F_2 seedlings for ~14 days in the growth chamber.
2. Harvest a small piece of leaf tissue from each seedling and incubate with 50 μL of the reconstitution buffer in each well of a 96-well plate. After >6 h incubation, apply the discharging solution to each well and monitor the bioluminescence signal. If the seedling is not expressing aequorin, no bioluminescence will be visible from the tissue. Select only those seedlings that show strong bioluminescence consistent with aequorin expression.
3. Harvest leaf tissue from the selected seedlings from **step 2** above and extract genome DNA. Use this DNA to genotype the plants by PCR using gene-specific primers and a T-DNA

left-border-specific primer (*see* **Note 2**). Select those plants that possess a homozygous T-DNA insertion and then grow to obtain additional seeds in the next generation.

4. In the F3 generation, homozygosity is confirmed by the same method described above (*see* **steps 2** and **3** of Subheading 3.1).

3.2 In Vivo Aequorin Reconstitution and Monitoring of $[Ca^{2+}]_{cyt}$ Responses

Once homozygous aequorin lines are successfully obtained in the various mutant backgrounds, they can be used for measurement of cytoplasmic calcium levels. First, reconstitute the aequorin–coelenterazine complex in vivo by incubating the seedlings in the reconstitution buffer. In a side by side manner, measure the calcium levels of the mutant plants in direct comparison to control experiments using non-transgenic wild-type plants. This is important to verify that aequorin bioluminescence is triggered by increased $[Ca^{2+}]_{cyt}$, and not due to reactive oxygen species (ROS)-triggered coelenterazine chemiluminescence (*see* **Note 8**).

1. Sow the seeds along a straight line on an MS solid medium in a petri dish.
2. After incubation at 4 °C for 3 days to synchronize seed germination, place the petri dish vertically in a plant growth chamber at 22 °C under lights.
3. Transfer a 5-day-grown seedling from the petri dish to an individual well of a 96-well microplate containing 50 μL of reconstitution buffer with coelenterazine and incubated overnight in the dark at room temperature.
4. After overnight incubation (*see* **Note 9**), place the 96-well plate under the camera and monitor the basal level to be stable before chemical applications.
5. Apply chemicals of your interest at the desired final concentrations (e.g., add 50 μL of double strength of the final concentrations), and then continue photon counting for a reasonable term (Fig. 2). We recommend the use of multichannel pipettes to simultaneously apply the chemical solutions to all wells.
6. At the end of each experiment, add an equal volume of the discharging solution (e.g., 100 μL) to discharge out remaining aequorin (*see* **Note 10**). Continue photon counting after discharge until the luminescence signal is well below the baseline level.

3.3 Calculation and Data Analysis: Converting Photon Counting Data into Ca^{2+} Concentration

Several procedures have been published to convert photon counts into $[Ca^{2+}]$ [19–22]. In our case, we use an equation adapted from Allen et al. [19] to determine in vivo $[Ca^{2+}]$ as follows:

1. Obtain the counting data as photons per second (L). In the case of measuring integrated $[Ca^{2+}]$ (Fig. 3), acquire the sum of the all photon counting data during the experiments.

2. Obtain total counting data for the entire sample over the course of the experiment until all the aequorin was discharged (L_{max}).
3. Input the data into the following equation: $[Ca^{2+}](nM) = \{[X^{1/3} + (X^{1/3} \times 55) - 1]/(1 - X^{1/3})\}/0.02$ (*see* **Note 11**), where X is the amount of light per second divided by the total light emitted after that time point until all the aequorin was discharged, i.e., $X = L/L_{max}$.
4. For all treatments, at least six replicates should be performed per group, and the individual $[Ca^{2+}]$ from repeat experiments averaged (Fig. 3).

4 Notes

1. In the examples shown here, we used transgenic aequorin lines in the wild-type Arabidopsis Col-0 background [10], in which 95–98 % of the aequorin is expressed in the cytoplasm [23]. There are other transgenic Arabidopsis lines expressing aequorin targeted to specific organelles, such as mitochondria [24], vacuolar microdomain [23], chloroplast [18, 25], and nucleus [26]. Aequorin can be fused with other reporter proteins such as green fluorescent protein (GFP) [27, 28], which can be used to localize the protein and analyze Ca^{2+} mobility in tissue or at a cellular level by bioluminescence resonance energy transfer.
2. Mutant lines are available at Arabidopsis Biological Resource center (https://abrc.osu.edu). In the examples shown here, we used *gpa1-4* (CS6534) and *agb1-2* (CS6536). Homozygous T-DNA insertions are selected by PCR using gene-specific primers [29, 30] and a T-DNA left border-specific primer (5′-GCGTGGACCGCTTGCTGCAACT-3′). Genomic DNA is isolated according to the protocol described by Edwards et al. [31].
3. Coelenterazine can also be dissolved in methanol, but never in organic solvents such as dimethylformamide and dimethylsulfoxide because coelenterazine spontaneously oxidizes. All the procedures using coelenterazine should be conducted under dim light. Prices of the chemical differ widely in companies although in our limited survey we found no differences in quality or purity. In addition to the native coelenterazine substrate, synthetic derivatives are available. For instance, *h-coelenterazine* and *cp*-coelenterazine have been used to demonstrate small changes in $[Ca^{2+}]_{cyt}$ in plants due to the increased Ca^{2+} sensitivity of aequorin.

4. The solution should be prepared immediately before use since coelenterazine is very sensitive to light and easily oxidizes. Once dissolved in aqueous solution, the chemical will start to be oxidized due to interaction with ambient oxygen. The best results are obtained with 1–10 μM of coelenterazine at final concentration.
5. White plates are commonly used for luminescent assays because of their reflective property; white plates reflect light within each well and will maximize the light output signal.
6. Recombinant aequorin emits ~5×10^{15} photons/mg protein [32]. In plant tissues, the actual levels of aequorin are relatively low: a few pg protein/mg fresh weight of tissue [33]. Therefore, very sensitive light counting equipment is required to detect the blue light. We used an image-intensified CCD camera. This is a very sensitive device since the signal-to-background ratio is improved by increasing the strength of the signal from the light sensors to detectors using microchannel plates. This device has an advantage of a quick scan speed to detect rapid changes in light emission in contrast to the cooled CCD camera.
7. It will be easier to screen the aequorin transgenic lines if the location of the aequorin transgene on the genome is determined in advance. We identified the insertion position of the aequorin used in our experiments by using TAIL-PCR. The gene is inserted at 4,341,822 bp on chromosome 1. This knowledge allows us to design primers to detect the transgene by PCR-base genotyping. Therefore, it would be difficult to establish useful mutant lines if the mutation were closely linked to the site of aequorin insertion on chromosome 1.
8. Coelenterazine is an effective antioxidant used as a chemiluminescent indicator for ROS production [34]. Indeed, in nature, coelenterazine have originally been used in ROS detoxification system by ancient marine organisms. Although light release from aequorin-bound coelenterazine is triggered by Ca^{2+}, any free coelenterazine can produce light when exposed to ROS. To avoid this problem, plants should not be stressed during the reconstitution step. Additionally, a control experiment is required using non-transgenic wild type plants in parallel and under identical condition.
9. Coelenterazine is a membrane-permeable molecule. Therefore, it is not necessary to use surfactants and vacuum infiltration to facilitate loading the molecule into tissue. However, different from the case of in vitro reconstitution (saturated after 6 h), aequorin activity continues to increase over a 24 h time period during in vivo reconstitution of 7-day-old Arabidopsis

seedlings [23]. The optimal incubation time for reconstitution is best obtained empirically in each lab, although for the example shown here we incubated the seedlings for 18–24 h.

10. To estimate the remaining aequorin after the experimental treatment, the tissues are immersed in a solution containing a massive amount of ethanol and Ca^{2+}. The ethanol destroys the cell membrane allowing Ca^{2+} penetration and discharge of the remaining aequorin. The discharging data is used for normalization of the results (*see* **steps 2** and **3** of Subheading 3.3).
11. The equation was adopted from a method based on the calibration curve of Allen et al. [19]: $L / L_{max} = \{(1 + K_R \times [Ca^{2+}]) / (1 + K_{TR} + K_R \times [Ca^{2+}])\}^3$, where K_R is the dissociation constant for the first Ca^{2+} to bind to aequorin, and K_{TR} is the binding constant of the second Ca^{2+} to bind to aequorin; $K_R = 2 \times 10^6$ M^{-1} and $K_{TR} = 55$ M^{-1}. The values are specific for the affinity between aequorin and coelenterazine derivatives, e.g., *cp*-coelenterazine: $K_R = 26 \times 10^6$ M^{-1} and $K_{TR} = 57$ M^{-1} [35].

Acknowledgments

We are grateful to Dr. Alan M.Jones (University of North Carolina, USA) for the Arabidopsis heterotrimeric G-protein mutants, and to Dr. Marc R.Knight (University of Oxford, UK) for transgenic Arabidopsis plants expressing aequorin. This work was supported by the US Department of Energy (grant no. DE-FG02-08ER15309) and by the Next-Generation BioGreen 21 Program, Systems and Synthetic Agrobiotech Center, Rural Development Administration, Republic of Korea (grant no. PJ009068).

References

1. Gilman AG (1987) G proteins: transducers of receptor-generated signals. Annu Rev Biochem 56:615
2. Sprang SR (1997) G protein mechanisms: insights from structural analysis. Annu Rev Biochem 66:639
3. Jones AM, Assmann SM (2004) Plants: the latest model system for G-protein research. EMBO Rep 5:572
4. Gelli A, Higgins VJ, Blumwald E (1997) Activation of plant plasma membrane Ca^{2+}-permeable channels by race-specific fungal elicitors. Plant Physiol 113:269
5. Ueguchi-Tanaka M et al (2000) Rice dwarf mutant d1, which is defective in the alpha subunit of the heterotrimeric G protein, affects gibberellin signal transduction. Proc Natl Acad Sci U S A 97:11638
6. Wu Y et al (2007) Heterotrimeric G-protein participation in Arabidopsis pollen germination through modulation of a plasmamembrane hyperpolarization-activated Ca^{2+}-permeable channel. New Phytol 176:550
7. Zhang W, Jeon BW, Assmann SM (2011) Heterotrimeric G-protein regulation of ROS signalling and calcium currents in Arabidopsis guard cells. J Exp Bot 62:2371
8. Shimomura O, Johnson FH, Saiga Y (1962) Extraction, purification and properties of aequorin, a bioluminescent protein from the luminous hydromedusan, Aequorea. J Cell Comp Physiol 59:223

9. Brini M et al (1995) Transfected aequorin in the measurement of cytosolic Ca^{2+} concentration ($[Ca^{2+}]_c$). A critical evaluation. J Biol Chem 270:9896
10. Knight MR, Campbell AK, Smith SM, Trewavas AJ (1991) Transgenic plant aequorin reports the effects of touch and cold-shock and elicitors on cytoplasmic calcium. Nature 352:524
11. Tanaka K, Swanson SJ, Gilroy S, Stacey G (2010) Extracellular nucleotides elicit cytosolic free calcium oscillations in Arabidopsis. Plant Physiol 154:705
12. Tanaka K, Gilroy S, Jones AM, Stacey G (2010) Extracellular ATP signaling in plants. Trends Cell Biol 20:601
13. Qi Z et al (2010) Ca^{2+} signaling by plant Arabidopsis thaliana Pep peptides depends on AtPepR1, a receptor with guanylyl cyclase activity, and cGMP-activated Ca^{2+} channels. Proc Natl Acad Sci U S A 107:21193
14. Ranf S et al (2012) Defense-related calcium signaling mutants uncovered via a quantitative high-throughput screen in *Arabidopsis thaliana*. Mol Plant 5:115
15. Wan J et al (2012) LYK4, a LysM receptor-like kinase, is important for chitin signaling and plant innate immunity in Arabidopsis. Plant Physiol 160:396
16. Baum G, Long JC, Jenkins GI, Trewavas AJ (1999) Stimulation of the blue light phototropic receptor NPH1 causes a transient increase in cytosolic Ca^{2+}. Proc Natl Acad Sci U S A 96:13554
17. Harada A, Sakai T, Okada K (2003) Phot1 and phot2 mediate blue light-induced transient increases in cytosolic Ca^{2+} differently in Arabidopsis leaves. Proc Natl Acad Sci U S A 100:8583
18. Nomura H et al (2012) Chloroplast-mediated activation of plant immune signalling in Arabidopsis. Nat Commun 3:926
19. Allen DG, Blinks JR, Prendergast FG (1977) Aequorin luminescence: relation of light emission to calcium concentration – a calcium-independent component. Science 195:996
20. Campbell AK (1988) Chemiluminescence – principles and applications in biology and medicine. VCH and Ellis Horwood Ltd., New York
21. Love J, Dodd AN, Webb AA (2004) Circadian and diurnal calcium oscillations encode photoperiodic information in Arabidopsis. Plant Cell 16:956
22. Torrecilla I, Leganes F, Bonilla I, Fernandez-Pinas F (2000) Use of recombinant aequorin to study calcium homeostasis and monitor calcium transients in response to heat and cold shock in cyanobacteria. Plant Physiol 123:161
23. Knight H, Trewavas AJ, Knight MR (1996) Cold calcium signaling in Arabidopsis involves two cellular pools and a change in calcium signature after acclimation. Plant Cell 8:489
24. Logan DC, Knight MR (2003) Mitochondrial and cytosolic calcium dynamics are differentially regulated in plants. Plant Physiol 133:21
25. Johnson CH et al (1995) Circadian oscillations of cytosolic and chloroplastic free calcium in plants. Science 269:1863
26. van Der Luit AH, Olivari C, Haley A, Knight MR, Trewavas AJ (1999) Distinct calcium signaling pathways regulate calmodulin gene expression in tobacco. Plant Physiol 121:705
27. Baubet V et al (2000) Chimeric green fluorescent protein-aequorin as bioluminescent Ca^{2+} reporters at the single-cell level. Proc Natl Acad Sci U S A 97:7260
28. Kiegle E, Moore CA, Haseloff J, Tester MA, Knight MR (2000) Cell-type-specific calcium responses to drought, salt and cold in the Arabidopsis root. Plant J 23:267
29. Jones AM, Ecker JR, Chen JG (2003) A reevaluation of the role of the heterotrimeric G protein in coupling light responses in Arabidopsis. Plant Physiol 131:1623
30. Ullah H et al (2003) The beta-subunit of the Arabidopsis G protein negatively regulates auxin-induced cell division and affects multiple developmental processes. Plant Cell 15:393
31. Edwards K, Johnstone C, Thompson C (1991) A simple and rapid method for the preparation of plant genomic DNA for PCR analysis. Nucleic Acids Res 19:1349
32. Shimomura O (1991) Preparation and handling of aequorin solutions for the measurement of cellular Ca^{2+}. Cell Calcium 12:635
33. Knight H, Knight MR (1995) Recombinant aequorin methods for intracellular calcium measurement in plants. In Galbraith DW, Bohnert HJ, Bourque DP (eds) Methods in Cell Biology 49:201–216
34. Dubuisson ML et al (2000) Antioxidative properties of natural coelenterazine and synthetic methyl coelenterazine in rat hepatocytes subjected to tert-butyl hydroperoxide-induced oxidative stress. Biochem Pharmacol 60:471
35. Shimomura O, Musicki B, Kishi Y, Inouye S (1993) Light-emitting properties of recombinant semi-synthetic aequorins and recombinant fluorescein-conjugated aequorin for measuring cellular calcium. Cell Calcium 14:373

Chapter 6

Methods for Analysis of Disease Resistance and the Defense Response in Arabidopsis

Guojing Li, Xiujuan Zhang, Dongli Wan, Shuqun Zhang, and Yiji Xia

Abstract

Many G proteins are involved in the plant defense responses against pathogen infection. With Arabidopsis as a model, this chapter describes the protocols commonly used for inoculating plants with *Pseudomonas syringae* (a bacterial pathogen) and *Botrytis cinerea* (a fungal pathogen) for analyzing disease resistance phenotypes caused by these pathogens. In addition, the procedures are provided for observation of the hypersensitive response triggered by avirulent strains of *P. syringae* and for analyzing pathogen-responsive gene expression and MAPK activation.

Key words Plant disease resistance, The hypersensitive response, *Pseudomonas syringae*, *Botrytis cinerea*, MAPK activity assay

1 Introduction

Plants employ complex mechanisms for protection against pathogen attack. A basal defense response is triggered once an invading pathogen is recognized by a host cell receptor that detects a microbe-associated molecular pattern (MAMP) [1, 2]. MAMPs are conserved molecules associated with certain types of microorganisms such as the flagellin protein from bacteria. The defense signaling pathway is mediated by the mitogen-activated protein kinase (MAPK) cascade and leads to activation of a number of defense-related genes [3]. On the other hand, pathogens evolved various mechanisms to suppress the host defense responses. For instance, *Pseudomonas syringae* secretes several dozens of proteins (called virulence effectors) into host cells to suppress the basal defense response [4], resulting in successful colonization. As a counteracting mechanism, plants have evolved various Resistance (R) proteins, each of which directly or indirectly recognizes one or multiple corresponding virulence effectors to trigger a strong defense response termed the hypersensitive response (HR). HR is associated with

Mark P. Running (ed.), *G Protein-Coupled Receptor Signaling in Plants: Methods and Protocols*, Methods in Molecular Biology, vol. 1043, DOI 10.1007/978-1-62703-532-3_6, © Springer Science+Business Media, LLC 2013

rapid death of the infected area and other defense mechanisms to prevent spread of invaded pathogens. A pathogen strain that can successfully colonize a host plant to cause disease is often called a virulence strain, whereas a strain that triggers HR through the R-effector recognition, thereby unable to cause disease, is termed an avirulence strain.

Many G proteins, including small G proteins, different subunits of heterotrimeric G proteins, and Extra-Large G Protein 2 (XLG2), have been found to play roles in plant–pathogen interactions [5–7]. Mutations in the genes encoding those G-proteins/subunits have been reported to alter disease resistance and/or the defense response including the MAPK cascade and defense gene induction. In this chapter, we describe the methods that are used to analyze disease resistance traits and the defense responses using Arabidopsis as a model. *P. syringae* is a biotrophic (or hemibiotrophic) pathogen that infects aerial parts of many economically important plant species such as tomato, beans, many cereal species, and woody species. Typical symptoms caused by *P. syringae* include cankers, leaf and fruit spots, and necrotic leaves. Several commonly used strains in Arabidopsis research belong to the pathovars *tomato* and *maculicola*. *Botrytis cinerea* is an airborne necrotrophic pathogen that also attacks a large number of plant species. It typically causes a soft rotting symptom of aerial plant parts and produce grey conidiophores and (macro) conidia [8]. These two pathogens have been used as models in studying Arabidopsis–pathogen interactions.

To determine whether a mutation causes alteration in disease resistance, we often infect plant leaves by using a syringe to hand-infiltrate a bacterial suspension, usually a virulent strain of *P. syringae* at a concentration of around 10^5 colony formation unit (cfu)/mL suspension. Disease symptoms can be observed by the naked eye. A more quantitative method for determining disease traits is by counting bacterial numbers in the infected leaves. To examine whether a mutation alters the hypersensitive response, leaves are infected by hand-infiltration of an avirulent strain of *P. syringae* at a higher concentration (around 10^7 cfu/mL), and hypersensitive cell death is observed by the naked eye by the symptom of collapse of the infected area within 5–24 h post-infection. *B. cinerea* is inoculated to plants often by using its spores and necrotic symptoms are recorded several days post infection. The inoculated plant tissues can also be used to study defense gene induction and pathogen-triggered MAPK activation. MAPK activation assay is performed using in-gel kinase activity assay described in this chapter or through Western blotting analysis using the antibodies raised against phospho-TEY that detect phosphorylated/activated MAPKs [9].

2 Materials

2.1 For Growing P. Syringae

1. King's medium B (*see* **Note 1**): to make 1 L of the K's B medium, add:

Proteose peptone #3 (Difco 0122)	20 g
Glycerol	10 mL
K_2HPO_4	1.5 g

   Adjust pH to 7.0–7.2 and make the volume to 1 L with demineralized water. For the K's B agar medium, add 15 g bacteriological agar for 1 L medium.

2. Rifampicin stock: Dissolve 50 mg of rifampicin in 1 mL of methanol, DMF, or DMSO to make a 50 mg/mL stock. Aliquot and store at −20 °C. For a working solution, the stock solution is added to the K's B medium to a final concentration of 50 μg/mL (*see* **Note 2**).
3. Kanamycin stock: Make a 50 mg/mL stock solution by dissolving 0.5 g of kanamycin disulfate salt into 10 mL of ddH_2O. Filter through a 0.22 μm filter to sterilize. Aliquot and store at −20 °C. Add the stock solution to the medium to bring the final concentration to 50 μg/mL (*see* **Note 3**).

   Rifampicin and kanamycin are added when autoclaved media cool to around 60 °C.
4. *P. syringae* strains:

   Virulent strains (e.g., *Pst* DC3000) or avirulent strains (e.g., *Pst AvrRpm1* or *Pst AvrRpt2*). *Pst AvrRpm1* and *Pst AvrRpt2* express and secrete the virulence effector AvrRpm1 and AvrRpt2, respectively, into plant cells. Most ecotypes of Arabidopsis (such as Col-0) express the corresponding R proteins that recognize AvrRpm1 and/or AvrRpt2 to trigger the hypersensitive response.

2.2 Pressure Infiltration of P. syringae

1. 1 mL needleless syringe.
2. A blunt-ended permanent marker.
3. 10 mM $MgCl_2$.
4. Arabidopsis plants: 4–6 week-old seedlings. Generally, plants are grown under a short day condition (e.g., 9 h light/15 h dark) so that the plants will remain at the vegetative stages with large leaves when they are infected.
5. Disposable 50 and 15 mL plastic tubes.

2.3 Inoculation with Botrytis cinerea

1. *Botrytis cinerea* (strain: DSM 4709) [10].
2. Hemocytometer.
3. Disposable 50 mL plastic tubes.
4. Miracloth.
5. Microscope.
6. Potato dextrose medium (Difco): For full-strength, dissolve 24 g potato dextrose medium in 1 L ddH_2O with 1.5 % agar. For half-strength, add 12 g/L with 1.5 % agar.
7. Half-strength MS medium:

MS salt (Sigma)	2.15 g
Sucrose	2.5 g
MES	0.25 g

Dissolve the chemicals in distilled water, adjust pH to 5.8 with KOH and bring the volume to 1 L. Autoclave under a liquid cycle for 30 min and store the medium at room temperature or in a cold room for future use.

2.4 MAPK Activity Assay

1. Protein extraction buffer:

HEPES	100 mM, pH 7.5
EDTA	5 mM
EGTA	5 mM
DTT	10 mM
Na_3VO_4	10 mM
NaF	10 mM
Glycerol	10 %
β-glycerophosphate	50 mM
Phenylmethylsulfonyl fluoride	1 mM
Antipain	5 μg/mL
Aprotinin	5 μg/mL
Leupeptin	5 μg/mL
Polyvinylpolypyrrolidone	7.5 %

The last four components are added just before use.

2. Washing buffer:

Tris	25 mM, pH 7.5
DTT	0.5 mM
Na_3VO_4	0.1 mM
NaF	5 mM
BSA	0.5 mg/mL
Triton X-100	0.1 % (v/v)

3. Renaturation buffer:

Tris	25 mM, pH 7.5
DTT	1 mM
Na_3VO_4	0.1 mM
NaF	5 mM

4. Reaction buffer:

Tris	25 mM, pH 7.5
EGTA	2 mM
$MgCl_2$	12 mM
DTT	1 mM
Na_3VO_4	0.1 mM
ATP	200 nM
Fresh γ-^{32}P-ATP (3,000 Ci/mmol)	50 μCi

5. Stop solution:

Trichloroacetic acid	5 % (w/v)
NaPPi	1 % (w/v)

6. SDS-polyacrylamide gel.

3 Methods

3.1 Inoculation of Plants with P. syringae by Pressure Infiltration

3.1.1 Analyzing Disease Susceptibility Phenotypes

Inoculum Preparation

1. Spread *P. syringae* onto the K's B agar medium with appropriate antibiotics such as rifampicin and kanamycin (*see* **Note 4**). Incubate the plate upside down in an incubator at 28 °C.
2. After growing for 24–48 h, the bacteria culture is visible from the surface of the plate. Shave the culture gently with an inoculating loop, and suspend the culture in sterile water or in 10 mM $MgCl_2$.
3. Measure the concentration of the bacterial suspension at 600 nm wavelength using a spectrophotometer.
4. Dilute the suspension to an appropriate concentration in sterile water or in 10 mM $MgCl_2$. Usually a concentration around 1×10^5 cfu/mL is recommended.

Pressure Infiltration

In addition to hand infiltration described below, other inoculation procedures can also be used to infect Arabidopsis plants with *P. syringae* [11].

1. Select fully or nearly fully expanded leaves and mark them with a blunt-ended permanent marker. Avoid using very old leaves that are at senescence.
2. Gently but firmly hold a needleless syringe against the abaxial side of a leaf and infiltrate approximately 10 μL of the bacterial suspension into the leaf area (Fig. 1). Either one half or both halves of a leaf can be inoculated. Avoid damaging the main veins. Cover the inoculated plants with lids for 1–2 days to maintain a high humidity.

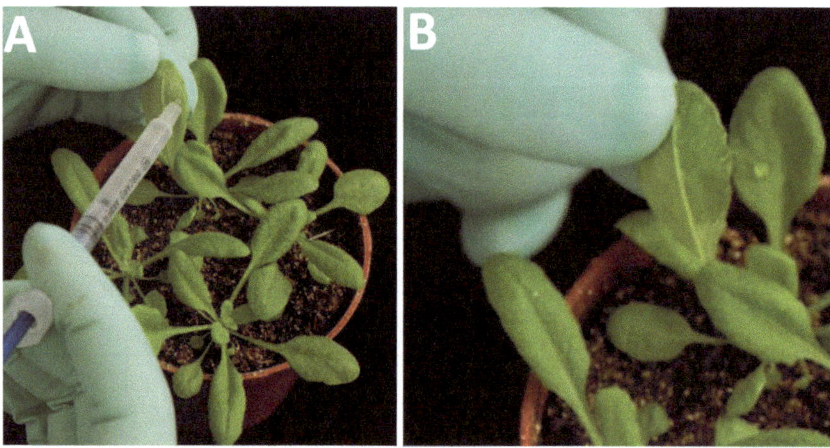

Fig. 1 Syringe infiltration of *Arabidopsis* leaves. (**a**) Gently but firmly hold a syringe against the leaf and the finger tip and infiltrate suspension into the leaf's intercellular space. (**b**) The syringe-infiltrated leaf with the infiltrated area appears water-soaked

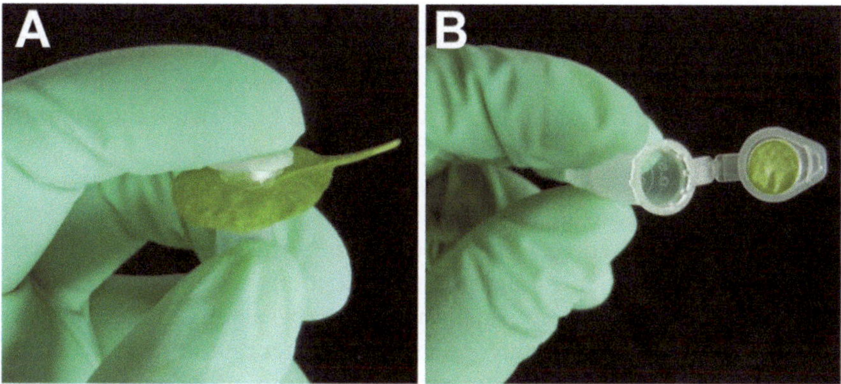

Fig. 2 Harvesting leaf discs with caps of 0.5 mL microcentrifuge tubes. (**a**) Cut a disc by closing the tube over the portion of the leaf to be sampled or by pressing the leaf against the cap using fingers. (**b**) a harvested leaf disc

Bacterial Counting

The infected area will display a necrotic symptom approximately 3–4 days post-inoculation. For more quantitative assay of the disease traits, bacterial numbers in the infected area are counted.

1. Leaf disks are cut from the infected areas at different time points (usually at 0, 1, 2, 3, and 4 days post-infection) with a cork borer (with an area of 0.25 cm^2 or smaller) or using caps of the 0.5 mL microcentrifuge tubes (Fig. 2). Four to eight leaf disks are pooled as one sample, and at least three replicates should be included for each time point.
2. Place the leaf disks for each sample into a 1.5 mL centrifuge tube. Add 200 μL sterile distilled water to each tube, and grind the sample thoroughly with a pestle.
3. Add 800 μL sterile distilled water to the samples. Mix and briefly spin down (for 15 s) at 425 ×g to precipitate pieces of tissue debris.
4. Make a series of 1:10 dilutions. Take a certain volume of the diluted suspension and spread on K's B agar plates (60-mm Petri dishes) that contain appropriate antibiotics to prevent contamination. An ideal density of bacterial colonies for accurate counting is 50–500 colonies/plate. For each sample, we usually prepare three plates with series of dilution which are estimated to generate 20, 200, and 2,000 colonies/plate, respectively. Based on our experience, each leaf disk (0.25 cm^2) might contain roughly 10^7 cfu at 3–4 days post-infection. Dilution series are made based on such estimation. However, bacterial growth rates are affected by many factors, including plant physiological states, growth conditions, ecotypes and genotypes, and bacterial strains. You need to figure out which dilution factors are suitable based on your experimental conditions.

5. Incubate the plates (upside down) at 28 °C for approximately 1–2 days and then count the colonies from each plate. Based on the dilution factors, calculate the cfu per cm^2 leaf disks. A bacterial growth curve can be drawn if the bacterial numbers are obtained at multiple time points.

3.1.2 Plant Inoculation for Inducing the Hypersensitive Response and for Analyzing Expression Profiling of Pathogen-Responsive Genes and MAPK Activation

An avirulent pathogen strain (such as *Pst AvrRpt2* or *Pst AvrRpm1*) is used to induce the hypersensitive response. Both virulent and avirulent strains induce distinct but highly overlapping set of genes; however, inoculation with an avirulent strain generally leads to a stronger induction of defense genes and is more often used for studying defense gene expression profiles. The procedures for inoculum preparation and pressure infiltration of bacterial suspension are similar to those described in the previous section. However, a higher level of an inoculum (often from 5×10^6 to 5×10^7 cfu/mL) is used. The time it takes for the visible hypersensitive response (collapse of inoculated areas) to appear depends on bacterial strains, levels of inoculums, and growth conditions. Generally it takes 5–24 h after inoculation before the symptom becomes visible to the naked eyes, as shown in Fig. 3. In addition to the observation by the naked eyes, more quantitative analysis of the hypersensitive cell death can be performed by analysis of ion leakage from cells undergoing HR [12, 13].

To study expression of pathogen-response genes or MAPK activation, leaves can also be inoculated as described above. Whole leaves are collected for RNA or protein extraction. Another set of

Fig. 3 The hypersensitive response (HR) in Arabidopsis leaves caused by *Pst AvrRpm1* inoculation. Leaves were infiltrated with 5×10^7 cfu/mL *Pst AvrRpm1*. (**a–b**) Arabidopsis plants before inoculation (**a**) and 6 h after inoculation (**b**). HR (collapse of the inoculated areas) became visible 6 h post-inoculation. (**c**) A close-up view of one of the inoculated leaves showing the HR symptom on the inoculated side

leaves are infiltrated with water (or 10 mM $MgCl_2$ if the bacterial culture is suspended in the $MgCl_2$ solution) as a negative control (mock inoculation). Changes in transcript levels can be detected within 15 min post-inoculation for some genes. Genes are induced or suppressed within a couple hours of pathogen inoculation are often considered as "early" pathogen-responsive genes, whereas those that change expression levels after 1 day post-infection are considered "late" pathogen-responsive genes. MAPK activation can be detected within 10 min post-inoculation.

3.2 B. cinerea Inoculation

3.2.1 Inoculating B. cinerea for Analyzing Disease Traits

1. *B. cinerea* is grown at 25 °C on the full-strength potato dextrose medium plates containing 1.5 % agar. It is maintained by sub-culturing every 4 weeks on the same medium.
2. For spore inoculum preparation, cut four small pieces of agar containing the *B. cinerea* culture and place them onto half-strength potato dextrose medium plates with 1.5 % agar and incubate at room temperature for about 12 days.
3. Scrape the spores with a stainless medicine spoon when the surface of hyphae shows visible dark color spores. Wash the spores with 10–15 mL half-strength MS medium (*see* **Note 5**).
4. Collect the spores into a 50 mL centrifuge tube. Shake vigorously or vortex to separate the spores with the hyphae.
5. Filter the spores through a 4-layer Miracloth and collect the filter-through with a new 50 mL centrifuge tube.
6. Centrifuge the filter-through at 2,900 × *g* for 15 min at room temperature to precipitate the spores.
7. Resuspend the spore pellet in an appropriate volume of half-strength MS medium.
8. Add the spore suspension to the center of the large middle square of a hemocytometer, 10 μL each side. Cover with a cover glass and let the spores settle down for 3 min.
9. Count the spore number of 5 squares under a microscope, and calculate the concentration as follows:

   Spore concentration per milliliter = total spore count in 5 squares × 50,000 × dilution factor.
10. Adjust the final concentration to about 5×10^5 spores/mL.
11. Grow *Arabidopsis* plants for 4–6 weeks in soil. Cover the plants with a humidity lid the day before infection.
12. Drop 10 μL of the spore suspension on the upside of each leaf surface.
13. Cover the inoculated plants with the humidity lid for 2 days under the normal plant growth condition.
14. Slightly lift/tilt the humidity lid to let the humidity drop gradually and maintain for another 1–2 days.

15. Score the disease symptoms 3 days after inoculation using the following index.
    (a) Have no obvious symptom.
    (b) Show necrosis spots and chlorosis.
    (c) Show significant chlorosis.
    (d) The fungal hyphae are visible.

3.2.2 Preparing B. cinerea-Inoculated Plants for Analysis of Defense Gene Expression

1. Prepare the spore inoculum as in the **steps 1–9** of Subheading 3.2.1.
2. *Arabidopsis* seedlings are grown in the 20-mL GC vials containing 6 mL half-strength MS medium. After growing at 22 °C under continuous light (about 70 μE/m^2/s) for 12–14 days, the seedlings are ready for inoculation.
3. Add the *B. cinerea* spore suspension into the GC vials with Arabidopsis seedlings to reach a final concentration of about 5×10^5 spores/vial.
4. The seedlings are then collected at various time points, weighed, quick-frozen in liquid nitrogen and stored at −80 °C until use for RNA extraction.

3.3 MAPK Activity Assay

1. Preparation of protein extracts:
   Ground 50–100 mg leaves (or other tissues) to fine powder in liquid nitrogen in 1.5 mL microcentrifuge tubes using plastic pestles. Then add 0.25 mL of the extraction buffer, and grind further to homogenate the tissues.
2. Centrifuge the extracts at 17,000 × *g* for 30 min and collect the supernatant. Measure the concentration of protein extracts using the Bio-Rad protein assay kit with BSA as a standard.
3. About 10 μg of protein is electrophoresed on 10 % SDS-polyacrylamide gels embedded with 0.25 mg/mL of myelin basic protein in the separating gel as a substrate for the kinase. Use prestained size markers (Bio-Rad) to calculate the sizes of kinases that will be detected.
4. After electrophoresis, wash the gel to remove SDS with the washing buffer three times (30 min each) at room temperature.
5. Renature kinases in the renaturation buffer at 4 °C. Change the buffer three times during the first 3 h of incubation, then leave the gel in the renaturation buffer overnight.
6. Incubate the gel at room temperature in 30 mL reaction buffer for 60 min.
7. Stop the reaction by putting the gel in the stop solution.
8. Remove unincorporated γ-^{32}P-ATP by washing the gel in the stop solution for at least 6 h. Change the stop solution five times during the wash period.

9. Dry the gel on Whatman 3MM paper with a gel dryer for about 1 h until the gel is completely dried.
10. Using a Phosphorimager or by exposing the gel to an X-ray film to quantify the relative kinase activities.

4 Notes

1. A low-salt Luria–Bertani (LB) medium can also be used instead of the K's B medium. The recipe is as follows:

Select peptone	10 g
Yeast extract	5 g
NaCl	5 g

Adjust the pH to 7.0.

2. Rifampicin can be stable for up to 2 years when stored at −20 °C and protected from light. Many laboratory strains of *P. syringae* are resistant to rifampicin. Rifampicin is added to prevent contamination. If a strain is not rifampicin-resistant, the antibiotic should not be added to the medium.
3. Kanamycin is used for strains that carry the Kan^R gene in a plasmid. Corresponding antibiotics should be used for strains that carry a different antibiotic-resistant gene.
4. Alternatively, the bacterial culture can be prepared by growing in a liquid medium with shaking at 200–250 rpm, 28 °C for about 12–24 h until it reaches mid to late log phase growth (OD_{600} = 0.6–1.0). Centrifuge the culture at 5,000 × *g* for 10 min to collect the bacteria cells and resuspend the pellet in sterile water or in 10 mM $MgCl_2$.
5. Cautions: since *B. cinerea* is an airborne pathogen, measures need to be taken to avoid spread of the spores during the experiments.

References

1. Glazebrook J (2005) Contrasting mechanisms of defense against biotrophic and necrotrophic pathogens. Annu Rev Phytopathol 43:205–227
2. Jones JDG, Dangl JL (2006) The plant immune system. Nature 444:323–329
3. Asai T, Tena G, Plotnikova J, Willmann MR, Chiu W-L, Gomez-Gomez L, Boller T, Ausubel FM, Sheen J (2002) MAP kinase signaling cascade in *Arabidopsis* innate immunity. Nature 415:977–983
4. Büttner D, He SY (2009) Type III protein secretion in plant pathogenic bacteria. Plant Physiol 150:1656–1664
5. Lieberherr D, Thao NP, Nakashima A, Umemura K, Kawasaki T, Shimamoto K (2005) A sphingolipid elicitor-inducible mitogen-activated protein kinase is regulated

by the small GTPase OsRac1 and heterotrimeric G-protein in rice. Plant Physiol 138:1644–1652
6. Zhu H, Li GJ, Ding L, Cui X, Berg H, Assmann SM, Xia Y (2009) Arabidopsis extra large G-protein 2 (XLG2) interacts with the Gbeta subunit of heterotrimeric G protein and functions in disease resistance. Mol Plant 2:513–525
7. Zhang W, He SY, Assmann SM (2008) The plant innate immunity response in stomatal guard cells invokes G-protein-dependent ion channel regulation. Plant J 56:984–996
8. Williamson B, Tudzynski B, Tudzynski P, van Kan JA (2007) *Botrytis cinerea*: the cause of grey mould disease. Mol Plant Pathol 8:561–580
9. Ahlfors R, Macioszek V, Rudd J, Brosché M, Schlichting R, Scheel D, Kangasjärvi J (2004) Stress hormone-independent activation and nuclear translocation of mitogen-activated protein kinases in *Arabidopsis thaliana* during ozone exposure. Plant J 40:512–522
10. Ren D, Liu Y, Yang KY, Han L, Mao G, Glazebrook J, Zhang S (2008) A fungal-responsive MAPK cascade regulates phytoalexin biosynthesis in Arabidopsis. Proc Natl Acad Sci USA 105:5638–5643
11. Katagiri F, Thilmony R, He SY (2002) The *Arabidopsis thaliana–Pseudomonas syringae* interaction. Arabidopsis Book 1:e0039. doi:10.1199/tab.0039
12. Mackey D, Holt BF III, Wiig A, Dangl JL (2002) RIN4 interacts with *Pseudomonas syringae* type III effector molecules and is required for RPM1-mediated resistance in Arabidopsis. Cell 108:743–754
13. Torres MA, Dangl JL, Jones JD (2002) Arabidopsis gp91^{phox} homologues AtrbohD and AtrbohF are required for accumulation of reactive oxygen intermediates in the plant defense response. Proc Natl Acad Sci USA 99:517–522

Chapter 7

Fusarium oxysporum Infection Assays in Arabidopsis

Yuri Trusov, David Chakravorty, and Jose Ramon Botella

Abstract

Increased susceptibility to *Fusarium oxysporum* is one of the most conspicuous characteristics of the Arabidopsis mutants lacking the heterotrimeric G protein β and γ1 subunits. The molecular mechanisms placing these G proteins in the plant innate immunity network are yet to be discovered. However, a method to test susceptibility to and disease progression of an important plant pathogen, such as *F. oxysporum*, is of central importance to many plant defense studies. The optimized protocol presented here allows the routine processing and analysis of symptom progression in young Arabidopsis soil-grown seedlings and yields highly reproducible results.

Key words *Fusarium oxysporum*, Hemi-biotrophic fungi, Arabidopsis, Plant pathogens

1 Introduction

Providing an appropriate response to environmental as well as internal developmental signals is vital for all living organisms. Signal processing typically depends on a number of molecules connected to each other and thereby forming specific signal transduction pathways. One such signal transduction pathway involves G-protein-coupled receptors (GPCRs) and heterotrimeric G proteins. Upon ligand recognition GPCRs activate heterotrimeric GTP-binding proteins which consist of three subunits: α, β and γ organized in a highly conserved complex. Activation induces a conformational change in Gα, catalyzing the exchange of GDP for GTP. Following the exchange, the heterotrimer dissociates into two functional elements, the Gα subunit and the Gβγ dimer. These two signaling elements interact with a number of downstream effectors. Intrinsic GTPase activity of Gα causes hydrolysis of GTP to GDP, which leads to re-association of the heterotrimer and its return to the pending state until next the signaling event [1, 2].

Since their discovery in plants [3–8] heterotrimeric G proteins have been implicated in a number of important cellular processes including plant defense or plant innate immunity [6, 9–11].

Mark P. Running (ed.), *G Protein-Coupled Receptor Signaling in Plants: Methods and Protocols*, Methods in Molecular Biology, vol. 1043, DOI 10.1007/978-1-62703-532-3_7, © Springer Science+Business Media, LLC 2013

Despite the fact that the important role of G proteins in defense against a number of pathogenic fungal species has been unambiguously established, the molecular mechanism has not yet been discovered. More studies in this direction are required and, therefore, methodological approaches should be presented in detail.

Fusarium oxysporum (f. sp. *conglutinans*) is a soilborne hemibiotrophic fungus that penetrates plants through the root tip, secondary root formation points, and wounds, to subsequently colonize the plant through the vascular system. Typical disease symptoms of Fusarium infection in Arabidopsis are the appearance of chlorosis starting from leaf veins and the retardation in plant growth, which ultimately results in death of the host [9, 12, 13].

From our experience *F. oxysporum* effects on Arabidopsis G protein mutants and wild type plants can vary greatly depending on experimental conditions, age of the plants and genetic background (ecotype). Therefore it is vital to optimize and standardize the assay conditions for the system of interest. Here we present a protocol optimized for Arabidopsis Columbia-0 and mutants: *gpa1-4*, *agb1-2*, *agg1-1c*, and *agg2-1*, lacking Gα, Gβ, Gγ1, and Gγ2 subunits, respectively.

2 Materials

1. *Fusarium oxysporum* culture.
2. Agar plates (1 % bacto-agar, pH 5.8).
3. Potato dextrose broth (PDB) (Fluka; Australia) medium (24 g/L); autoclaved.
4. Schott bottle (1 L), autoclaved.
5. Miracloth or a plastic sieve.
6. Healthy *Arabidopsis thaliana* seedlings (10–15 days old).
7. Soil is based on University of California potting mix C, fertilizer II, with readily available nitrogen plus moderate reserve of nitrogen, pH 6.5 [14].
8. Pots 40–50 mm in diameter, 50–70 mm tall.
9. Growing trays 80 mm × 140 mm × 50 mm.
10. Incubation trays (300 mm × 350 mm) with clear plastic lids.

3 Methods

3.1 Arabidopsis Seedling Preparation

Plant approximately 50–100 seeds of each line to be analyzed and appropriate controls in small (40–50 mm in diameter) pots. Seven days after germination immerse the small pot in water and carefully remove seedlings from the soil. Transplant about 30–40 seedlings into growing trays (80 mm × 140 mm × 50 mm). Make sure that seedlings are separated from each other by at least 10 mm. Grow plants for another 5–7 days (*see* **Note 1**).

3.2 Cultivating *F. oxysporum*

Although some *F. oxysporum* strains do have antibiotic resistance (up to 100 μg/mL of streptomycin and up to 15 μg/mL of tetracycline [15]) we recommend conducting all work with the culture under sterile conditions until inoculation day. Short hyphae and spores of *F. oxysporum* can be stored for months blotted on sterile 3 M paper at 4 °C. Seven days before inoculation place paper block with spores on a plain agar plate. Incubate at room temperature for 2–4 days (usually 2 days is enough) until fungal hyphae are visible (*see* **Note 2**). Prepare and autoclave potato dextrose broth (PDB) medium (24 g/L in water). We found that a 200 mL culture in a 1-L Schott bottle (can be used to hold 200–300 mL of culture) produces enough spores for inoculation of 100 plants. Inoculation of the liquid culture is performed by excising two or three blocks (about 5 mm × 5 mm) of agar containing hyphae using a sterile scalpel blade and transferring them to the bottle with PDB medium equilibrated to room temperature. Incubate the culture for 3–4 days on a rotary shaker at 110 rpm and 28 °C (*see* **Note 3**). Filter the culture through four layers of Miracloth or a plastic sieve into an inoculation tray. In the literature it is often recommended to spin down the spore solution, remove the PDB medium and resuspend spores in water for inoculation. We, however, found that inoculation in PDB medium or in water produce identical results if antibiotics were not supplemented with the PDB. Spore concentration in the inoculum affects symptom development considerably. Therefore, calculate spore concentration using a hemocytometer and dilute (if necessary) to 10^6 spores/mL with distilled water.

3.3 Inoculation

Immerse an Arabidopsis tray containing 30–40 seedlings into water, carefully remove one by one 20 seedlings and place them into a container with clean distilled water until all 20 seedlings have been collected. Shake seedlings gently in the water to remove excess soil. Blot seedlings on tissue or blotting paper for 1–2 s and immerse them into inoculation suspension for at least 30 s. Longer inoculation time does not affect the results significantly, therefore, it is possible to accumulate several seedlings in the inoculation tray before planting. Using forceps make a hole in the soil and plant the infected seedlings one by one into the incubation tray, separated by at least 15 mm. It is better to use two forceps: one for making holes and another to handle the seedlings. Cover the tray with a clear plastic lid and incubate the infected seedlings at 26–29 °C, depending on experimental design. The rest of the seedlings should be replanted into a separate tray and used as mock inoculation control. This control is necessary to ensure that handling of seedlings during inoculation, which causes unavoidable leaf and root damage, and subsequent growth conditions are not responsible for leaf chlorosis and general plant health. Excessive damage during manipulation will compromise final results.

3.4 Evaluation of the Resistance/ Susceptibility Levels

Levels of resistance/susceptibility are usually expressed as estimates/values of disease symptoms displayed by infected plants. In the *F. oxysporum*—Arabidopsis system the most pronounced and easily measurable symptoms are leaf chlorosis starting with yellowing leaf veins, reduction of rosette growth, wilting of the inflorescence (if it develops) and eventually plant decay (although recovery is also observed in some cases). First symptoms manifested as leaf veins yellowing become clearly visible at about 5–6 days post inoculation in hypersensitive *agb1-2* mutants and 6–7 days in wild type Col-0. Depending on plant growth and infection conditions as well as the resistance levels of the infected plants, the tested plants initially display leaf yellowing and eventually could decay or recover from the disease. For this reason it is necessary to evaluate the disease progression at different stages of symptom development; in some cases we will score progression daily.

The disease progression or symptom development can be evaluated in several ways. The simplest and most subjective method is an expert estimate. It can be used for quick and rough evaluation and is especially useful when processing many plants. The idea is to inspect all infected plants visually, determine the range of disease progress (usually based on general plant "sickness"), subdivide the range on several classes ascribing a value for each class (for instance, "healthy"—0; "a little sick"—1; "very sick"—2; and "dead"—3) and finally evaluate every plant in respect to this criterion. Calculated averages for tested lines could be presented in comparison with the control line. This way, however, relies heavily on the operator's experience and is highly prone to human error.

Counting number of leaves with yellow veins is usually found to be a reliable and effective way of disease evaluation, although is not completely impervious to human error. Both the disease progression and chlorosis development are gradual processes and, therefore, it is difficult to be certain about the status of a particular leaf. We recommend counting only leaves with clearly developed vein chlorosis or when veins turned a definite yellow color. Resolution of this method could be further increased if it involves also counting total number of leaves per plant and presenting data as average ratio of yellow/green or yellow/total leaves. Statistical significance could be estimated by Student's *t*-test.

F. oxysporum inhibits plant growth, by blocking vascular system and thus decreasing water supply to the leaves. This inhibition results in smaller leaf blades and shorter peduncles, which collectively could be easily measured and presented as rosette diameter. Importantly, rosette size is a variable quantitative trait affected by numerous genes including *GPA1* and *AGB1* encoding Gα and Gβ subunits respectively. Therefore reduction in rosette growth caused by *F. oxysporum* infection requires measurements from infected and mock inoculated plants to take into account the effect of the mutation. This possibility should be considered for any study. Therefore an average rosette size for mock inoculated plants should be determined first. Then, for every individually infected plant, the ratio rosette size/average rosette size (mock inoculated) should be calculated and averaged for the line. Comparison could be presented as average reduction. Statistical significance could be estimated by Student's *t*-test.

Depending on the specific growth conditions during post inoculation, infected plants of the same genotype could either die or recover. This can then be an additional criterion for evaluation of disease resistance levels: percentage of decayed or recovered plants. From our experience evaluations of disease symptoms at early stage (number of yellow vein leaves) and late stage (percentage of decayed plants) do not always correlate [10]. Chi-square or Anova statistics could be used for estimating significance of the differences.

4 Notes

1. Plants should be as healthy and uniform as possible. Stratify seeds for at least 2 days after sowing to ensure simultaneous germination. During replanting, select seedlings of equal size.
2. Fusarium hyphae are thin and white; therefore, inspect the plate using light, illuminating the plate from the opposite side.
3. Longer incubation (more than 4 days) will result in spore germination, while in shorter incubations the culture does not produce sufficient spores. Also note that if the *F. oxysporum* culture appears "milky," especially after 1 or 2 days of incubation, it indicates bacterial contamination and it should be discarded. By day 3–4 it is normal to observe large number of small "spheres" of hyphae.

References

1. Engelhardt S, Rochais F (2007) G proteins: more than transducers of receptor-generated signals? Circ Res 100:1109–1111
2. Gilman AG (1987) G proteins: transducers of receptor-generated signals. Annu Rev Biochem 56:615–649
3. Ma H, Yanofsky MF, Meyerowitz EM (1990) Molecular cloning and characterization of GPA1, a G protein α subunit gene from *Arabidopsis thaliana*. Proc Natl Acad Sci USA 87:3821–3825
4. Mason MG, Botella JR (2000) Completing the heterotrimer: isolation and characterization of an *Arabidopsis thaliana* G protein γ-subunit cDNA. Proc Natl Acad Sci USA 97:14784–14788
5. Mason MG, Botella JR (2001) Isolation of a novel G-protein γ-subunit from *Arabidopsis thaliana* and its interaction with Gb. Biochim Biophys Acta 1520:147–153
6. Suharsono U, Fujisawa Y, Kawasaki T, Iwasaki Y, Satoh H, Shimamoto K (2002)

The heterotrimeric G protein a subunit acts upstream of the small GTPase Rac in disease resistance of rice. Proc Natl Acad Sci USA 99:13307–13312

7. Weiss CA, Garnaat CW, Mukai K, Hu Y, Ma H (1994) Isolation of cDNAs encoding guanine nucleotide-binding protein β-subunit homologues from maize (ZGB1) and Arabidopsis (AGB1). Proc Natl Acad Sci USA 91: 9554–9558
8. Chakravorty D, Trusov Y, Zhang W, Sheahan MB, Acharya BW, Mccurdy DW, Assmann SM, Botella JR (2011) A highly atypical heterotrimeric G protein γ subunit is involved in guard cell K^+ channel regulation and morphological development in *Arabidopsis thaliana*. Plant J 67(5):840–851
9. Trusov Y, Rookes JE, Chakravorty D, Armour D, Schenk PM, Botella JR (2006) Heterotrimeric G-proteins facilitate Arabidopsis resistance to necrotrophic pathogens and are involved in jasmonate signaling. Plant Physiol 140:210–220
10. Trusov Y, Sewelam N, Rookes JE, Kunkel M, Nowak E, Schenk PM, Botella JR (2009) Heterotrimeric G proteins-mediated resistance to necrotrophic pathogens includes mechanisms independent of salicylic acid-, jasmonic acid/ethylene- and abscisic acid-mediated defense signaling. Plant J 58:69–81
11. Llorente F, Alonso-Blanco C, Sanchez-Rodriguez C, Jorda L, Molina A (2005) ERECTA receptor-like kinase and heterotrimeric G protein from Arabidopsis are required for resistance to the necrotrophic fungus *Plectosphaerella cucumerina*. Plant J 43:165–180
12. Agrios GN (2005) Plant pathology. Academic, New York
13. Trusov Y, Jorda L, Molina A, Botella JR (eds) (2010) G Proteins and plant innate immunity. Springer, Netherlands
14. Baker KF (1957) The UC system for producing healthy container-grown plants. Agricultural Experiment Station, University of California, Berkeley, CA
15. Campbell EJ, Schenk PM, Kazan K, Penninckx IA, Anderson JP, Maclean DJ, Cammue BP, Ebert PR, Manners JM (2003) Pathogen-responsive expression of a putative ATP-binding cassette transporter gene conferring resistance to the diterpenoid sclareol is regulated by multiple defense signaling pathways in Arabidopsis. Plant Physiol 133:1272–1284

Chapter 8

Analysis of Unfolded Protein Response in Arabidopsis

Yani Chen and Federica Brandizzi

Abstract

The unfolded protein response (UPR) is fundamental for development and adaption in eukaryotic cells. *Arabidopsis* has become one of the best model systems to uncover conserved mechanisms of the UPR in multicellular eukaryotes as well as organism-specific regulation of the UPR in plants. Monitoring the UPR *in planta* is an elemental approach to identifying regulatory components and to revealing molecular mechanisms of the plant UPR. In this chapter, we provide protocols for the induction and analyses of plant UPR at a molecular level in *Arabidopsis*. Three kinds of ER stress treatment methods and quantitation of the plant UPR activation are described here.

Key words Unfolded protein response, UPR, Endoplasmic reticulum stress, Tunicamycin, Protein folding, *Arabidopsis*

1 Introduction

The unfolded protein response (UPR) is a collection of signaling pathways aiming to maintain endoplasmic reticulum (ER) protein folding homeostasis in eukaryotic cells [1, 2]. There is approximately one-third of total protein folded and modified in the ER. Environmental or physiological factors that cause an imbalance between demand and capability of ER protein folding lead to ER stress. To relieve the ER stress, increase in the ER protein folding ability is one of the most instant and central regulation in the UPR. Diverse stimuli from exogenous or endogenous signals trigger the activation of the UPR. To experimentally examine the UPR, chemicals disturbing the ER protein folding homeostasis are applied to induce the UPR. One of the most frequently used UPR inducers is a glycosylation inhibitor, tunicamycin (Tm). The majority of secretory proteins are glycosylated in the ER as the glycosylation is crucial for protein structure formation and for protein targeting to cellular compartments. As Tm blocks the first step of N-linked glycosylation, it can efficiently lead to accumulation of unfolded protein in the ER lumen and therefore activate the UPR [3–6].

Mark P. Running (ed.), *G Protein-Coupled Receptor Signaling in Plants: Methods and Protocols*, Methods in Molecular Biology, vol. 1043, DOI 10.1007/978-1-62703-532-3_8, © Springer Science+Business Media, LLC 2013

To observe the plant UPR at different growth stages, we describe three experimental approaches to perform ER stress treatment. To investigate long-term ER stress tolerance, seeds are directly germinated on medium containing a relatively low concentration of Tm. Tm can also been infiltrated into leaves to monitor the UPR specifically on ground tissues. Finally, to examine the early outputs of the plant UPR, a short-term Tm treatment using a liquid method is conducted to observe a more instant and direct response to ER stress.

To cope with dynamic ER protein folding demands, the UPR adjusts the transcription of genes function in assembling protein structure, degrading mis-folded protein, and determining cell fates [7, 8]. Hence, the upregulation of well-established UPR target genes, such as *BiP3* in Arabidopsis [9], is considered a molecular indicator of UPR activation. To introduce the quantitative method of reading UPR outputs, real-time reverse transcription polymerase chain reaction (qRT-PCR) analysis of UPR target genes induction is included in this chapter.

2 Materials

1. Basic reagents and equipment for plant sterile tissue culture and RNA work handling.
2. Plant growth medium: Linsmaier and Skoog (LS) with Buffer and Sucrose (Caisson LSP04); Phytagel (Sigma P8169).
3. Growth chamber: temperature set to 21 °C, 16 h light/8 h dark cycle, 100 mEinstein/m^2 s, and 65 % humidity.
4. Tunicamycin (Sigma T7765).
5. Dimethyl sulfoxide (DMSO) solvent.
6. 1 ml needleless syringes.
7. Liquid nitrogen.
8. RNeasy plant mini kit (Qiagen 74904).
9. RNase-Free DNase Set (Qiagen 79254).
10. SuperScript® VILO™ Master Mix (Invitrogen 11755500).
11. Reagents for qRT-PCR: MicroAmp® Fast optical 96-well reaction plate (ABI 4346936); optical adhesive cover (ABI 4311971); FAST SYBR Master Mix (ABI 4385612).

3 Methods

3.1 Germination Under the ER Stress

To examine the tolerant ability of plants in coping with different intensities of ER stress, seeds are directly germinated on medium containing Tm concentrations ranging from 10 to 50 ng/ml. Comparison of phenotype between wild-type plants and mutants of interest can reveal whether the mutants display over-sensitive or resistant growth phenotype under ER stress conditions. The Tm infiltration assay enables the observation of the plant UPR using adult plants.

1. Sterilize seeds and store at 4 °C for 2 days (*see* **Note 1**).
2. Prepare ½ LS with 0.4 % Phytagel medium.
3. Autoclave the ½ LS medium on liquid cycle program for 25–40 min.
4. Dissolve Tm powder in DMSO to prepare 10 mg/ml Tm stock solution (*see* **Note 2**).
5. Prepare 10, 20, 30, 40, and 50 μg/ml Tm stock solutions by 1,000, 500, 333, 250, and 200× dilution of 10 mg/ml Tm stock solution respectively using ½ LS liquid medium.
6. Cool the autoclaved ½ LS medium to 50 °C.
7. Add 10, 20, 30, 40, and 50 μg/ml Tm stock solutions respectively to cooled ½ LS medium (50 °C) by 1,000× dilution to make ½ LS medium containing 10, 20, 30, 40, and 50 ng/ml Tm (*see* **Note 3**).
8. Swirl to mix and use a pipette to pour equal amount of Tm-containing medium per plate in the sterile tissue culture hood. Prepare Tm-containing medium freshly right before the Tm germination assay (*see* **Note 4**).
9. For Mock control, the same preparation procedure is carried out with the exception of replacing the Tm in the ½ LS medium with 0.0005 % DMSO.
10. Germinate *Arabidopsis* seeds on ½ LS medium containing 0.0005 % DMSO, 10, 20, 30, 40, and 50 ng/ml Tm. Place a single seed on the medium in an equally spaced manner (*see* **Note 5**). Perform the germination at least in triplicate with minimal three individual plates for each Tm concentration and mock control.
11. Grow the plants under these conditions: 21 °C, 16 h light/8 h dark cycle, 100 mEinstein/m^2 s, and 65 % humidity.
12. Observe the growth phenotype 7–14 days after germination (*see* **Note 6**).
13. Using Col-0 ecotype wild-type Arabidopsis, the plants shows more pronounced growth defects starting from 30 ng/ml Tm (*see* Fig. 1).

3.2 Tm Infiltration into Leaf Tissues

1. Dissolve Tm powder in DMSO to prepare 10 mg/ml Tm stock solution (*see* **Note 2**).
2. Prepare 15 μg/ml Tm working solution by 666× dilution of 10 mg/ml Tm stock solution using ½ LS liquid medium. Prepare Tm-containing medium freshly right before the Tm infiltration assay (*see* **Note 4**).

Fig. 1 Wild-type Col-0 plants were germinated on ½ LS medium containing DMSO, 10, 20, 30, 40, or 50 ng/ml Tm for 2 weeks

3. Use a needleless syringe to infiltrate ½ LS liquid medium containing 15 μg/ml Tm into abaxial sides of 5-week-old rosette leaves (*see* **Note 7**).
4. For Mock control, the same treatment procedure is performed with the exception of replacing the Tm in the ½ LS liquid medium with 0.0015 % DMSO (*see* **Note 8**).
5. Observe the leaves phenotype 1–4 days after infiltration.

3.3 Short Period of ER Stress Treatment

While the ER stress tolerance assay can examine whether mutants of interest show a plant phenotype under ER stress, even if the mutants display a comparably visible plant phenotype to wild-type plants, it is possible that the defects of UPR in mutants of interest do not reflect on the plant growth morphology. For instance, a mutant of *AtbZIP60* shows compromised UPR activation phenotype at a molecular level but displays a similar tolerant plant phenotype when germinated under ER stress [10, 11]. To verify whether genes of interest are involved in the UPR, short-term ER stress treatment coupled with analyses of UPR target genes induction are performed to monitor the UPR at a molecular level.

1. Sterilize seeds and store at 4 °C for 2 days (*see* **Note 1**).
2. Germinate seeds in vertical plates for 10 days. Medium: ½ LS with 0.4 % Phytagel. Place ten seeds evenly spaced per small round plate (100×15 mm) or square plate. Seal the bottom part of plates with parafilm and the upper part of plates with 3M surgical tape (*see* Fig. 2 and **Note 9**).
3. Dissolve Tm powder in DMSO to prepare 10 mg/ml Tm stock solution (*see* **Note 2**).
4. Prepare 5 μg/ml Tm-containing medium by 2,000× dilution of 10 mg/ml Tm stock solution using ½ LS liquid medium. Prepare Tm-containing medium freshly right before the Tm treatment (*see* **Note 2**).

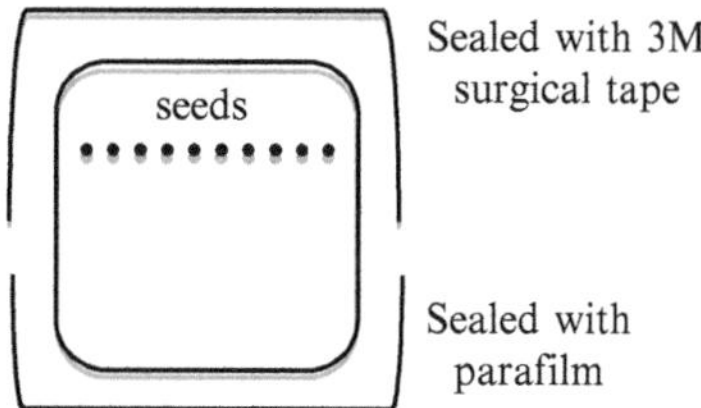

Fig. 2 The 3M surgical tapes and parafilm are used respectively to seal the upper and bottom part of vertical plates

5. Gently transfer 10-day-old vertically grown seedlings to 5 μg/ml Tm-containing medium for an appropriate time period (*see* **Notes 10** and **11**).
6. Collect 10–20 individual Tm-treated seedlings per biological replicate using liquid nitrogen (*see* **Notes 12** and **13**).
7. For Mock control, the same treatment procedure is performed with the exception of replacing the Tm in the ½ LS liquid medium with 0.05 % DMSO.

3.4 Quantitative Measurement of UPR Activation

The regulation of UPR target genes transcription is one of the major outputs of the plant UPR. Hence, measurement of UPR target genes induction under ER stress is a classical method to quantify the plant UPR activation.

1. Extract RNA from Tm-treated seedlings using an RNeasy plant mini kit and RNase-Free DNase Set.
2. Synthesize CDNA from RNA using a SuperScript® VILO™ Master Mix.
3. Perform qRT-PCR with SYBR Green detection in triplicate using the Applied Biosystems 7500 fast real-time PCR system. The primer sequence of UPR target genes is listed in Table 1 [11].
4. Analyze Data by the DDCT method.

4 Notes

1. The quality of seed stock is extremely important for ER stress related assays. Using seeds freshly harvested from healthy plants is one of key points to get dependable and consistent results.
2. Aliquot Tm stock solution (10 mg/ml) into relatively small amount and store in a −20 °C freezer. Avoid freezing and thawing.
3. High temperature destabilizes Tm.

Table 1
Primers of UPR target genes for the Applied Biosystems 7500 fast real-time PCR system

Primers	Sequence (5′–3′)	Gene
BiP1/2-qP For	ccaccggccccaagag	AT5G28540/AT5G42020
BiP1/2-qP Rev	ggcgtccacttcgaatgtg	AT5G28540/AT5G42020
BiP3-qP For	aaccgcgagcttggaaaat	AT1G09080
BiP3-qP Rev	tcccctgggtgcaggaa	AT1G09080
AtERdj3A-qP For	tcaagtggtggtggtttcaact	AT3G0890
AtERdj3A-qP Rev	cccaccgcccatattttg	AT3G0890
AtERdj3B-qP For	gaggaggcggcatgaatatg	AT3G62600
AtERdj3B-qP Rev	ccatcgaacctccaccaaaa	AT3G62600
PDI6-qP For	cgaagtggctttgtcattcca	AT1G77510
PDI6-qP Rev	gcggttgcgtccaatttt	AT1G77510
PDI9-qP For	ggccctgttgaagtgactgaa	AT2G32920
PDI9-qP Rev	cagcagaaccacacttcttttcc	AT2G32920
CNX1-qP For	gtgtcctcgtcgccattgt	AT5G61790
CNX1-qP Rev	ttgccaccaaagataagcttga	AT5G61790
CRT1-qP For	gatcaagaaggaggtcccatgt	AT1G56340
CRT1-qP Rev	gacggaggacgaaggtgtaca	AT1G56340
AtERdj2A-qP For	tgggcttgtaggcgctctt	AT1G79940
AtERdj2A-qP Rev	aacccaatagttttcctccttgtg	AT1G79940
AtERdj2B-qP For	tgaaacgtcccaatggactca	AT4G21180
AtERdj2B-qP Rev	cctctttgtggaaaggaaagtaagg	AT4G21180
AtP58^{IPK}-qP For	gcgttatagtgatgccctcgat	AT5G03160
AtP58^{IPK}-qP Rev	gaaagcgcagggtctgctt	AT5G03160

4. Tm-containing medium is unstable if it is not freshly prepared.
5. For a fair comparison, the distance between seeds should be consistent. Square petri dish with Grid (Fisher 08-757-11A) is useful as a single seed can be placed on the center of each small square area on the plates (*see* Fig. 1).
6. The growth phenotype can be more obvious at a relatively early stage (within 1 week) or vice versa. To detect stage-specific phenotypes, the growth of Tm-treated seedlings should be observed every day during the assay.

7. To fairly compare the ER tolerance between wild-type and mutant plants, choose the same stage, size, and condition of leaves for both varieties of plants.
8. The infiltration process needs be performed carefully and should not lead to any damage of plants. Leaves infiltrated with DMSO (Mock control) should appear comparable to leaves without infiltration after 1 day.
9. To allow proper ventilation, do not wrap plates completely with parafilm.
10. Select well-grown and unstressed seedlings as well as similar growth morphology for all plants.
11. Using 10-day-old seedlings coupled with a qRT-PCR system, the induction of UPR target genes can be detected starting from 0.5 h. Prolonged treatment is not recommended using this liquid system.
12. Using more seedlings per biological sample can reduce the standard deviation of fold change of UPR target genes induction between biological replicates.
13. Sample collection should be done carefully and timely to avoid additional stress before seedlings are frozen by liquid nitrogen.

Acknowledgments

This study was supported by grants from the National Institutes of Health (R01 GM101038-01), Chemical Sciences, Geosciences and Biosciences Division, Office of Basic Energy Sciences, Office of Science, U.S. DOE (DE-FG02-91ER20021), NASA (NNX12AN71G) and the National Science Foundation (MCB 0948584 and MCB1243792).

References

1. Kozutsumi Y, Segal M, Normington K, Gething MJ, Sambrook J (1988) The presence of malfolded proteins in the endoplasmic-reticulum signals the induction of glucose-regulated proteins. Nature 332:462–464
2. Back SH, Schroder M, Lee K, Zhang KZ, Kaufman RJ (2005) ER stress signaling by regulated splicing: IRE1/HAC1/XBP1. Methods 35:395–416
3. Takatsuk A, Arima K, Tamura G (1971) Tunicamycin, a new antibiotic.1. Isolation and characterization of tunicamycin. J Antibiot 24:215–223
4. Heifetz A, Keenan RW, Elbein AD (1979) Mechanism of action of tunicamycin on the UDP-GlcNAC-dolichyl-phosphate GlcNAC-1-phosphate transferase. Biochemistry 18:2186–2192
5. Keller RK, Boon DY, Crum FC (1979) N-acetylglucosamine-1-phosphate transferase from hen oviduct—solubilization, characterization, and inhibition by tunicamycin. Biochemistry 18:3946–3952
6. Brandish PE, Kimura K, Inukai M, Southgate R, Lonsdale JT, Bugg TDH (1996) Modes of action of tunicamycin, liposidomycin B, and mureidomycin A: inhibition of phospho-N-acetylmuramyl-pentapeptide translocase from Escherichia coli. Antimicrob Agents Chemother 40:1640–1644

7. Acosta-Alvear D, Zhou Y, Blais A, Tsikitis M, Lents NH, Arias C, Lennon CJ, Kluger Y, Dynlacht BD (2007) XBP1 controls diverse cell type- and condition-specific transcriptional regulatory networks. Mol Cell 27:53–66
8. Cox JS, Walter P (1996) A novel mechanism for regulating activity of a transcription factor that controls the unfolded protein response. Cell 87:391–404
9. Iwata Y, Koizumi N (2005) An Arabidopsis transcription factor, AtbZIP60, regulates the endoplasmic reticulum stress response in a manner unique to plants. Proc Natl Acad Sci U S A 102:5280–5285
10. Lu DP, Christopher DA (2008) Endoplasmic reticulum stress activates the expression of a sub-group of protein disulfide isomerase genes and AtbZIP60 modulates the response in Arabidopsis thaliana. Mol Genet Genomics 280:199–210
11. Chen YN, Brandizzi F (2012) AtIRE1A/AtIRE1B and AGB1 independently control two essential unfolded protein response pathways in Arabidopsis. Plant J 69:266–277

Chapter 9

Functional Analysis of Heterotrimeric G Proteins in Chloroplast Development in Arabidopsis

Wenjuan Wu and Jirong Huang

Abstract

Functional analysis of G-proteins has been extensively carried out using their over-expressing lines and knockout mutants in plants. Since α subunit exists in an active or inactive form, overexpressing α subunit does not mean that G-protein signaling pathways are activated in the transgenic lines. Ectopic expression of the constitutively active form of the α subunit will magnify a role of G-protein signaling pathways in plant growth and development, and ultimately yield phenotypes. Here, we describe the method to study the function of G-proteins in chloroplast development using the constitutively active form of the α subunit in Arabidopsis.

Key words Heterotrimeric G-proteins, Constitutively active GPA1, Chloroplast development, Variegation, THF1, Arabidopsis

1 Introduction

In eukaryotes, heterotrimeric G-proteins are called molecular switches that are turned on by G-protein coupled receptors (GPCR) while turned off by regulators of G-protein signaling (RGS). G-proteins play a role in multiple plant developmental and physiological processes, such as seed germination, leaf shape, pathogen infection, sugar perception and stomatal development and movement. Currently, the yeast two-hybrid system and genetic screening for suppressors of G-protein mutants are mostly used for dissection of G-protein signaling pathways, particularly for identification of downstream effectors, in plants. For example, in Arabidopsis several G-protein α subunit (GPA1) interacting proteins, such as a cupin domain-containing protein (PRN1), prephenate dehydratase 1 (PD1), and thylakoid membrane formation 1 (THF1) have been identified [1–4], whereas a Golgi-localized hexose transporter SGB1 (Suppressor of G-protein Beta1) and N-myc downregulated-like1 (NDL) are shown to genetically and physically interact to the β subunit (AGB1), respectively [5, 6].

Mark P. Running (ed.), *G Protein-Coupled Receptor Signaling in Plants: Methods and Protocols*, Methods in Molecular Biology, vol. 1043, DOI 10.1007/978-1-62703-532-3_9, © Springer Science+Business Media, LLC 2013

However, these approaches are limited in systematically dissecting a G-protein signaling pathway that specifically regulates a phenotype.

In the previous study, thylakoid formation1 (THF1) was identified to be an interacting protein for GPA1 [4]. Knockout of THF1 leads to a phenotype in leaf variegation, the severity of which is significantly affected by developmental and environmental conditions. This phenotype provides us a good opportunity to test whether G-proteins are involved in the process of chloroplast development. To test this hypothesis, G-protein signaling is switched on by knocking out RGS1 and transforming the constitutively active form of GPA1 (cGPA1) in the genetic background of *thf1* [7]. Our results showed that overexpression of *cGPA1* rescued the leaf variegation phenotype of *thf1*, and rectify transcriptional levels of many mis-regulated genes in *thf1*. Interestingly, chloroplast development in another leaf variegation mutant *var2/ftsh2* can be also significantly rescued by ectopic expression of *cGPA1*. This overlooked function of G-proteins provides new insight into our understanding of the integrative signaling network, which dynamically regulates chloroplast development and function in response to both intracellular and extracellular signals. More importantly, the transgenic plant provides a good system to identify novel components in the established genetic pathway for G-proteins to regulate chloroplast development.

2 Materials

2.1 Plant Materials

The *Arabidopsis thaliana* ecotype Columbia-0 was used. *thf1* and *var2* mutants used were previously described [5, 8].

2.2 Strains and Plasmids

1. One Shot TOP10 chemically competent cells (Invitrogen).
2. DH5α *E. coli* chemically competent cells.
3. GV3101 Agrobacterium chemically competent cells.
4. pENTR/SD/D-TOPO (Invitrogen).
5. pGWB2 (Research Institute of Molecular Genetics, Shimane University, Japan).

2.3 Solutions and Reagents

1. 1/2 MS media for plant growth: Add 2.15 g Murashige and Skoog basal salt mixtures (MS, Sigma) and 10 g sucrose to 800 mL of water and stir to dissolve, adjust pH to 5.7 with 1 M NaOH and bring to 1 L with water. Add 7 g of Phyto agar (Duchefa Biochemie) to the liquid medium before autoclaving. After the medium cools to about 50 °C, add the required selection marker and pour the medium into petri dishes (*see* **Note 1**).
2. LB media: Add 10 g Tryptone, 5 g Yeast extract, and 10 g NaCl to 950 mL water, adjust pH to 7.0 with 1 M NaOH.

Adjust the volume to 1 L with water. Autoclave the media. For solid media, add 15 g Bacto Agar before autoclaving.

3. Antibiotics: Kanamycin (Amresco), Hygromycin (Invitrogen), Rifampicin (Sigma), Gentamycin (Amresco).
4. Infiltration medium for Agrobacterium-mediated transformation: Add 50 g of sucrose, 2.2 g of MS salt, and 0.5 g of MES to 800 mL of water, adjust pH to 5.7 using 1 M NaOH and bring to 1 L with water. Add 200 μL of silwet77 to 1 L of the media (*see* **Note 2**).
5. pENTR™ directional TOPO® cloning kits (Invitrogen, K2420-20).
6. Gateway® LR Clonase Enzyme Mix Kits (Invitrogen, 11791-043).
7. GeneTailor™ Site-Directed Mutagenesis System (Invitrogen, 12397-014).
8. QIAprep Spin Miniprep Kit (Qiagen, 27106).
9. RNAgents® Denaturing Solution (Promega, Z5651).
10. TPS extraction buffer: Add 5 mL 2 M KCl, 1 mL 1 M Tris–HCl pH 8.0, 1 mL 100 mM EDTA pH 8.0, and 3 mL water to prepare 10 mL TPS.
11. 70 % ethanol.

3 Methods

3.1 Plant Growth

Stratify seeds at 4 °C for 4 days, and then sow onto soil in a 22 °C growth chamber with an 8 h light/16 h dark cycle at light intensity of 100 μmol photons/m^2/s. Alternatively, seeds are surface-sterilized and planted onto 1/2 MS media. Six-day-old seedlings are transplanted to soil.

3.2 Plasmid Constructions

Plasmids are constructed using the Gateway cloning system (Invitrogen, http://www.invitrogen.com).

3.2.1 TOPO Cloning

1. The entire open-reading frame of *GPA1* is amplified by PCR from a cDNA library with the primers 5′-CACCATGGGC TTACTCTGCAGTAG-3′ and 5′-TCATAAAAGGCCAGCC TCCAGTA-3′ (*see* **Note 3**).
2. Purify the amplified fragment of *GPA1* from the PCR product and estimate the content of the amplified fragment using agarose gel electrophoresis.
3. Set up the TOPO cloning reaction. For optimal results, use a 0.5:1 to 2:1 molar ratio of PCR product to TOPO vector. Usually, add 0.5–4 μL fresh PCR product, 1 μL salt solution, TOPO vector 1 μL and add water to 6 μL (*see* **Note 4**).

4. Mix gently and spin down, incubate overnight at room temperature.
5. Transform the TOPO cloning reaction to One Shot POT10 Chemically Competent *E. coli*.
6. Isolate plasmids from positive colonies, and confirm the sequence of *GPA1* by sequencing.

3.2.2 Make the Constitutively Active Form of GPA1$^{(Q222L)}$ by Site-Directed Mutagenesis

The constitutively active form of GPA1$^{(Q222L)}$ is generated with the GeneTailor Site-directed mutagenesis system (Invitrogen, http://www.invitrogen.com).

1. Dilute the kit-supplied 200× SAM solution to 10× SAM (*see* **Note 5**).
2. For Methylation Reaction, mix up 100 ng plasmid DNA, 1.6 μL methylation buffer, 1.6 μL 10× SAM, 1 μL (10 U/μL) DNA methylase and bring to 16 μL with water. A single reaction provides enough methylated plasmid for up to eight mutagenesis reactions. Then Incubate the reagents at 37 °C for 1 h (*see* **Note 6**).
3. Site-directed mutagenesis with Platinum *Taq* DNA polymerase high fidelity. Prepare the 50 μL reaction mixture as followings: 5 μL 10× high fidelity PCR Buffer, 1.5 μL 10 mM dNTP, 1 μL 50 mM $MgSO_4$, 1.5 μL 10 μM Primers, 2–5 μL Methylated DNA, 0.2–0.5 μL Platinum *Taq* high fidelity (5 U/μL), and water. The primers used for site-directed mutagenesis are 5′-GATTGTTTGACGTGGGTGGACTGAGAAATGAGAG-3′ and 5′-TCCACCCACGTCAAACAATCGGTACACTTC-3′ (*see* **Note 7**).
4. Check 10–20 μL of the product on a 1 % agarose gel and transform mutagenesis reaction mixture to DH5α Competent cells.
5. Confirm the mutation site of *GPA1* by sequencing.

3.2.3 LR Reaction

The fragments are then recombined into the pGWB2 destination vector using LR Clonase Enzyme Mix Kits according to the manufacturer's instruction.

1. Mix 1–7 μL entry vector (50–150 ng), 1 μL 150 ng destination vector, 2 μL LR clonase II, and water to a final volume of 8 μL. For optimal results, use a 1:1 molar ratio of the entry vector to the destination vector.
2. Incubate the reaction at room temperature overnight.
3. Add 2 μL of the Proteinase K solution and incubate samples at 37 °C for 10 min to terminate the reaction.
4. Transform the LR mixture to DH5α competent cells.
5. Isolate the plasmids from the positive clones using the QIAprep Spin Miniprep Kit according to the manufacturer's instruction.

3.3 Plasmid Transformation into Agrobacteria GV3101

1. Add 50–100 ng of the GPA1$^{(Q222L)}$-pGWB2 plasmid into GV3101 chemically competent cells and mix gently.
2. Incubate on ice for 30 min.
3. Deep freeze the cells in the liquid N_2 and then heat-shock the cells for 3 min at 37 °C without shaking immediately.
4. Transfer the tubes to ice for 5 min.
5. Add 1 mL of room temperature LB liquid media and shake the tube (200 rpm) at 28 °C for 1 h.
6. Spread the cells on a pre-warmed selective plate containing Kanamycin (50 µg/mL), Hygromycin (50 µg/mL), Rifampicin (50 µg/mL), Gentamycin (50 µg/mL) and incubate for 2–3 days at 28 °C.

3.4 Plasmid Transformation into Arabidopsis Plants

Transform the plasmid into Arabidopsis plants of *thf1* or *var2* by the flower-dipping method [9].

1. Inoculate the transformed Agrobacteria in 5 mL of LB liquid media containing antibiotics, and shake (200 rpm) at 28 °C for 24 h.
2. Add 5 mL of Agrobacteria to 250 mL LB medium containing antibiotics and incubate Agrobacteria overnight with shaking (200 rpm) at 28 °C.
3. When the Agrobacteria grow to about 1 of OD_{600}, collect the Agrobacteria and resuspend in 250 mL of the infiltration medium.
4. Dip the full-blown flowers into the Agrobacteria solution for 5 min, cover the plants with cling wrap, and put them in the dark overnight.
5. Continue to grow the plants in the greenhouse and collect the seeds after maturation.

3.5 Transgenic Plant Screening and Confirmation

1. Surface-sterilized seeds are stratified at 4 °C for 4 days, and then sown on 1/2 MS media supplemented with kanamycin (50 µg/mL) for transgenic plant screening.
2. Transfer the transgenic seedlings to soil and grow continuously.
3. Harvest seeds from transgenic plants individually.
4. The single insertion of T-DNA can be determined by the segregation ratio of plants resistant to kanamycin to those not resistant to kanamycin in the T2 generation. The T-DNA insertion is confirmed by PCR using genomic DNA or cDNA isolated from transgenic plants. PCR products amplified from genomic DNA with gene-specific primers 5′-ACCGATTGTTT GACGTGGGTGGACT-3′ and 5′-CCGTGTTCTGGTAAT ATAACTCCTC-3′. The fragment amplified from the transgene cGPA1 is shorter than that from genomic GPA1 since an

intron is included in the fragment of genomic GPA1. Since the point mutation in cGPA1 introduces a restriction enzyme site of *Dde*I, transcripts of the transgene can be confirmed by the different patterns of *Dde*I-digested fragments, which are amplified from cDNA of Wild-type and transgenic cGPA1. PCR products amplified from cDNA of cGPA1 with gene-specific primers 5′-TTATATTCCAACTAAGGAGGATGTAC-3′ and 5′-CCGTGTTCTGGTAATATAACTCCTC-3′ can be cut by *Dde*I and generate three bands with 353 and 119 bp, whereas PCR products amplified from endogenous GPA1 cDNA cannot be digested by *Dde*I.

3.6 Total RNA Extraction

Total RNA was isolated using the RNAgents total RNA isolation system (Promega, http://www.promega.com) in accordance with the manufacturer's instructions (*see* **Note 8**).

3.7 DNA Isolation

1. Cut a leaf from transgenic plants and put it into a 1.5 mL tube and add 300 μL TPS extraction buffer.
2. Grind the tissues using sticks or TissueLyserII (QIAGEN).
3. Incubate the tubes in 75 °C water bath for 20 min.
4. Centrifuge for 10 min at 15,300 × *g*.
5. Pipette the suspension to a new tube and add the same volume of isopropyl alcohol.
6. Centrifuge for 15 min at 1,300 × *g*.
7. Remove the supernatant and add 700 μL of 70 % ethanol to the tube.
8. Centrifuge for 10 min at 15,300 × *g* and remove the supernatant.
9. Dry the DNA and dissolve in 20–50 μL of water.

4 Notes

1. If you put 1 L 1/2 MS media in several bottles, separate plant agar in average to each bottle proportional to the volume.
2. Please stir entirely after adding silwet77.
3. Be sure to include the 4 bp sequences (CACC) necessary for directional cloning on the 5′ end of the forward primer. Use a thermostable, proofreading DNA polymerase to produce the blunt-end PCR product.
4. For pENTR™ TOPO vectors, using 1–5 ng of a 1 kb PCR product or 5–10 ng of a 2 kb product in a TOPO cloning reaction generally results in a suitable number of colonies. Too much or too few PCR product would reduce the cloning efficiency.

5. 10× SAM is not stable and will lose activity within a few hours after preparation. Do not use 10× SAM if it is more than a few hours old.
6. The methylation reaction can be stored at −20 °C. The methylated DNA will remain stable at −20 °C for up to 3 months.
7. Usually, use 2 μL of methylation mixture (containing 12.5 ng of methylated plasmid) per 50 μL of mutagenesis reaction as a starting point. If the bands are faint, up to 5 μL may be required. Using more than 5 μL of methylation mix may decrease yield.
8. Wear groves all the time when manipulating with RNA to avoid contamination.

References

1. Lapik YR, Kaufman LS (2003) The Arabidopsis cupin domain protein AtPirin1 interacts with the G protein a-subunit GPA1 and regulates seed germination and early seedling development. Plant Cell 15:1578–1590
2. Zhao J, Wang X (2004) Arabidopsis phospholipase Da 1 interacts with the heterotrimeric G-protein a-subunit through a motif analogous to the DRY motif in G-protein-coupled receptors. J Biol Chem 279:794–800
3. Warpeha KM, Upadhyay S, Yeh J, Adamiak J, Hawkins SI, Lapik YR, Anderson MB, Lee BS, Kaufman LS (2007) The GCR1, GPA1, PRN1, NF-Y signal chain mediates both blue light and abscisic acid responses in Arabidopsis. Plant Physiol 143:1590–1600
4. Huang J, Taylor JP, Chen J, Uhrig JF, Schnell DJ, Nakagawa T, Korth KL, Jones AM (2006) The plastid protein thylakoid formation 1 and the plasma membrane G-protein GPA1 interact in a novel sugar-signaling mechanism in Arabidopsis. Plant Cell 18:1226–1238
5. Wang Q, Sullivan RW, Kight A, Henry RL, Huang J, Jones AM, Korth KL (2004) Deletion of the chloroplast-localized thylakoid formation1 gene product in Arabidopsis leads to deficient thylakoid formation and variegated leaves. Plant Physiol 136:3594–3604
6. Mudgil Y, Uhrig JF, Zhou J, Temple B, Jiang K, Jones AM (2009) Arabidopsis N-MYC DOWNREGULATED-LIKE1, a positive regulator of auxin transport in a G protein-mediated pathway. Plant Cell 21:3591–3609
7. Zhang L, Wei Q, Wu W, Cheng Y, Hu G, Hu F, Sun Y, Zhu Y, Sakamoto W, Huang J (2009) Activation of the heterotrimeric G protein alpha-subunit GPA1 suppresses the ftsh-mediated inhibition of chloroplast development in Arabidopsis. Plant J 58:1041–1053
8. Chen M, Choi Y, Voytas DF, Rodermel SR (2000) Mutations in the Arabidopsis *VAR2* locus cause leaf variegation due to the loss of a chloroplast FtsH protease. Plant J 22: 303–313
9. Clough SJ, Bent AF (1998) Floral dip: a simplified method for Agrobacterium-mediated transformation of Arabidopsis thaliana. Plant J 16:735–743

Chapter 10

G Protein Signaling in UV Protection: Methods for Understanding the Signals in Young Etiolated Seedlings

Danielle A. Orozco-Nunnelly, Lon S. Kaufman, and Katherine M. Warpeha

Abstract

A seed is competent to respond to light soon after imbibition. A new developmental program begins in or on the ground where the young seedling may be exposed to heat, cold, drought, flooding (anoxia), salts, varying levels of visible light, and the topic of this paper, ultraviolet radiation. Herein what is described is a method for growing and maintaining seedlings, then methods of UV irradiation in order to measure discrete effects of UV wavelengths in signal transduction, very early in seedling development. The physiological response to an abiotic signal is partly dependent on the developmental state of the plant. Dark-grown seedlings of plant species possess young leaves or leaf primordia in a "suspended" state of development whereby exposure to sunlight, visible and UV, is required to initiate the leaf developmental program, including development of etioplasts or proplastids into fully functioning chloroplasts. In order for us to understand the initial and persisting effects of UV in seedlings, we "delay" light-induced development by carrying out all experiments in complete darkness between days 0 (seed) and day 7 (Arabidopsis). In this case, the UV regulation of a simple signaling pathway in Arabidopsis, G protein signaling in UV protection and acclimation early in development, is investigated with the use of several mutants and easily score-able phenotypes.

Key words Ultraviolet, G-protein, G-protein-coupled receptor, Etiolated, Seedling, Complete darkness

1 Introduction

Even though seedlings can experience abiotic signals from day 1, many experiments on abiotic signals in the past have been on older plants (2–6 weeks old), either in the lab or in the field. Older plants have already experienced many stimuli, resulting in development and acclimation that include many gene expression, structural, biochemical changes. This makes the study of the actual importance of UV reception and signaling mechanism(s) in development more complex, and the specific study of UV indeed would be confounded by other environmental signals. A seed is competent to respond to light 48–56 h after imbibition [1].

Mark P. Running (ed.), *G Protein-Coupled Receptor Signaling in Plants: Methods and Protocols*, Methods in Molecular Biology, vol. 1043, DOI 10.1007/978-1-62703-532-3_10, © Springer Science+Business Media, LLC 2013

A new developmental program begins in or on the ground where the young seedling may be exposed to heat, cold, drought, flooding (anoxia), salts, varying levels of visible light, and the topic of this paper, ultraviolet radiation.

Phenylpropanoids are derived from the amino acid phenylalanine (Phe), and many of these organic compounds are directly involved in the plant's response to abiotic signals. Phenylpropanoids are critical to UV protection, including the UV-screening flavonoids, like quercetin, simple phenolics, such as the hydroxycinnamic acids, and waxy/fat-containing materials that occur in association with the plant's cuticle. Quercetin and other UV-screening pigments found in seeds vary with regard to both composition and concentration among different plant species (reviewed in ref. 2). The synthesis of UV-protective materials can occur as a developmental process, in response to UV-A or UV-B, perhaps in temporal anticipation of higher levels of UV-B, or to stress-inducing levels of UV-B [3–7]. Phenylpropanoids in general may build the plant's early defense mechanisms, where adequate induction of the phenylpropanoid pathway appears to be important in the first 3 weeks post-germination, influencing the plant's ability to respond to later stresses (reviewed in ref. 6). The level of Phe present in a germinating seed may be critical for the ability to acclimate to new environmental conditions. Hence, to understand the initial effects of UV reception and signaling, we study the early period after germination of the seed.

Herein what is described is a method for growing and maintaining seedlings, then methods of UV irradiation in order to measure discrete effects of UV wavelengths in signal transduction, very early in seedling development. The physiological response to an abiotic signal is partly dependent on the developmental state of the plant. Dark-grown seedlings of many dicotyledonous plant species (in particular) possess young leaves or leaf primordial in a "suspended" state of development whereby exposure to sunlight, visible and UV, is required to initiate the leaf developmental program. Among the changes induced include leaf expansion and development of immature plastids (etioplasts; proplastids) into fully functioning chloroplasts. In order for us to understand the initial and persisting effects of UV in seedlings, we "delay" light-induced development by carrying out all experiments in complete darkness between days 0 (seed) and day 7 (Arabidopsis) (*see* Fig. 1). In this case, the UV regulation of a simple signaling pathway in Arabidopsis, G protein signaling in UV protection and acclimation early in development, is investigated with the use of several mutants and easily score-able phenotypes. With regard to G-protein involvement, all experiments use mutants of GPA1 and GCR1 as a means of control to demonstrate involvement in G-protein signaling.

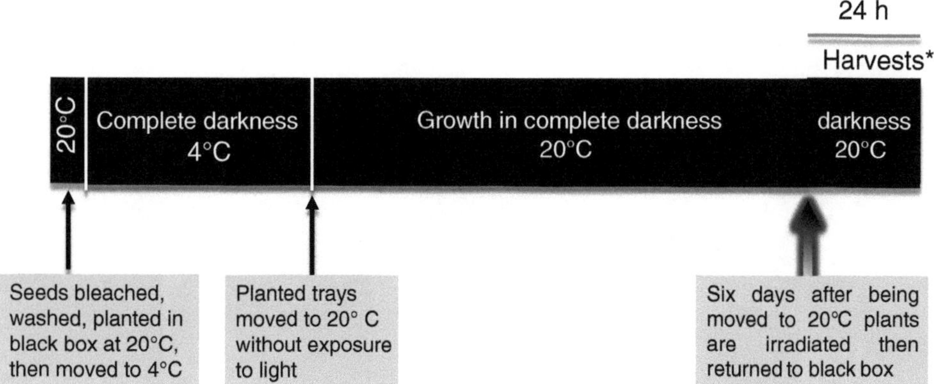

Fig. 1 The process of growing and maintaining plants for early UV effects on growth and development. Seedlings are not exposed to any irradiation at any time except during intended irradiation. Irradiation of any particular wavelength is set according to total photons being delivered at 10^4 μmol/m^2 or less, in <20 min. (*Asterisk*) Harvests can occur at any time from 0 to 24 h after irradiation period (*see* **Note 1**)

2 Materials

2.1 Planting Rooms/Areas

1. Completely light-tight growth room (can be incubator room or adapted lab room) (*see* **Note 2**).

   All seams have to be covered (Black opaque flexible plastic, sealed by black duct tape).

   A photography dark room is only suitable if the red safe lights are permanently off.
2. Clean-air hood advisable in room.
3. Chemical hood for UV sources (some give off ozone/fumes).
4. All areas surface sterilized (bleach wipes; ethanol; 20 % bleach; Fig. 2).

2.2 Portable Equipment

1. Constructing containers for plants.

   Source of black Plexiglas 5–6 mm thick with approximate dimensions 16″ × 13″ with a depth of 10″ to accommodate three tiers (3 × 3 = 9 per tier) of phytatrays.

   Boxes must be constructed where all seams are completely sealed (test by flashlight taped in inside of box in a completely dark room).

   The ends that meet the lid must be a beveled tight-fitting lid (made of black Plexiglas).
2. Fluorescent light box adapted as green safelight.

   A standard fluorescent lightbox (48″ × 12″) using tube light fittings.

   Only two Bulbs should be fitted for reduced fluence.

   Bulb type: Sylvania gold F40/GO.

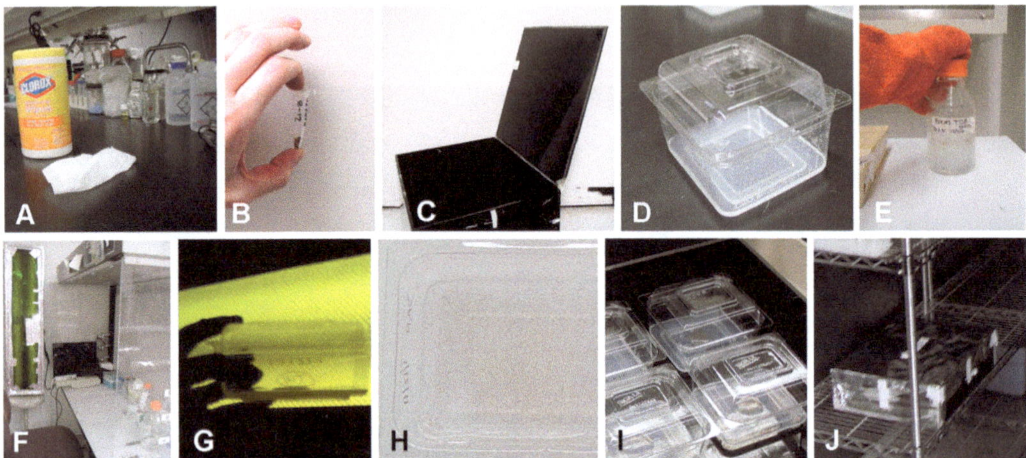

Fig. 2 Process of preparation for planting for dark-grown experiments. (**a**) Sterilize area, remove dust before experiment. (**b**) Measure out seed. (**c**) Use light-tight black box (also surface sterilize). (**d**) Preprepare phytatrays. (**e**) Prepare Top Agarose. (**f**) Dark room area with safe light. (**g**) The dim green safe light. (**h**) Appearance of seeds after pouring in Top Agarose. (**i**) Finished planting in the black box. (**j**) Tape black box, label, then foil to cover lid fully, then tape again to secure foil to box; placement in 4 °C for 48 h (then box moved to 20 °C)

The plastic cover is wrapped by two sheets of green cellulose-acetate theatrical gel.

On the outside of the cover, green Plexiglas 3 mm thick (peak emission of 518 nm).

The Plexiglas is cut to size to fit within the frame.

Assembly wrapped on all sides with black duct tape.

Final seals are aluminum Nash tape or flexitape.

Output should be measured with LI-COR LI-250 photometer using a PAR sensor (LI-COR, Lincoln, NE) and maximum 0.1 μmol/m^2 s at 12 in. from housing.

3. UV irradiation source.

UV-B and discrete UV-A wavelengths

Lamp designed to deliver UV-B and UV-A range at wavelengths 300, 305, 311, 317, 325, 332, and 368 nm (assembled by the UVBMRN, Colorado State University, Fort Collins, CO; [7]).

Light is redirected by a mirror that reflects the UV radiation.

Near-infrared (NIR) and infrared (IR) pass through an exhaust port.

Condensing lenses collect the reflected UV light and define the beam.

Beam passes through one of seven narrow band filters of type (UVMFRSR).

Narrow band-pass filters (Barr Associates, Westford, MA, USA) have a 2 nm FWHM at: 300, 305, 311, 317, 325, 332 and 368 nm.

Short irradiations (<20 min) selected to avoid high fluence responses and reciprocity failure, keeping fluence the same regardless of wavelength.

Machine mounted on stand to deliver an 8×9.5 cm square dose area (bottom of a phytatray) from above.

UV-protective equipment is required: face-shield, lab coat, gloves.

General UV-C and UV-A

Handheld UV lamp.

Dual band output: 254 nm (UV-C); UV-A broadband with 320–380 nm, peak ~366 nm.

(UV-P model UVGL-58 Mineralight lamp UVP, Upland, CA, USA).

UV-protective equipment is required: face-shield, lab coat, vinyl gloves.

2.3 Seed Stocks

1. Arabidopsis seed stocks of T-DNA insertion mutants were obtained from TAIR [8]. For G-protein studies, mutants of GCR1 (At1G48270); GPA1 (At2G26300), AGB (At4G34460), AGG1 (At3G63420), AGG2 (At3G22942) and AGG3 (At5g20635) are essential to compare to any other mutants tested.
2. Other seed stocks of crops (cultivars, etc.) are obtained from state or national depositories.

2.4 Planting Materials (All Chemicals Molecular Biology or Plant Cell Culture Grade; All Plastics Sterile)

1. Sigma gamma-irradiated phytatrays (Phytatray I or Phytatray II).
2. MS media.
3. MES.
4. 5 M KOH.
5. Top agarose (0.5× MS/MES pH 5.8; 0.8 % low melt agarose).
6. Standard molecular biology grade agarose.
7. Lab glassware (500 mL to 2.0 L Erlenmeyer flasks and bottles).
8. 15 mL sterile plastic tubes with volume gradations.
9. Standard bleach.
10. Autoclaved deionized water.
11. 25 or 2.5 % Triton-X 100 in sterile water.
12. Arabidopsis seed.

13. Parafilm squares cut to 1 in.
14. Extra heavy duty aluminum foil.
15. Autoclaved potting soil.
16. Nitrile gloves and hot gloves.
17. Bleach wipes.
18. Squares of aluminum foil that have been autoclaved.
19. Kitchen spatulas.
20. Small weigh boats.
21. 2 in. squares of wax paper.
22. Microbalance spatulas.
23. Buffers for extraction (HEPES or Phosphate or Tris depending on experiment; [14, 15]).

2.5 Lab Equipment

1. pH meter.
2. Autoclaves.
3. Magnetic stirrer.
4. Balances (pan and analytical).
5. Small water bath at 50 °C.
6. Timer (non-light-emitting).

2.6 Tools for Phenotypic Analysis

1. Camera.
2. Deconvoluting microscopy.
3. Dissecting scope/stereomicroscope.
4. SEM and TEM.

3 Methods

3.1 Making 0.5× MS/MES pH 5.8 Plates

1. In a 2.0 L Erlenmeyer flask, 4.33 g MS is weighed out into 1.5 L of ultrapure water under stirring.
2. 0.5 g MES measured out and added to Erlenmeyer flask as in **step 1**.
3. When all material dissolved, solution is adjusted to pH 5.8 with 5 M KOH at 20 °C then brought up to ~1,990 mL (*see* **Note 3**).
4. ~195 mL is poured into an autoclavable glass bottle and labeled "Top Agarose" whereby low melt agarose molecular grade is added to final 0.8 %, bottle capped with threaded cap and adjusted to 200 mL with ultrapure water.
5. The original solution is split into two 2.0 L Erlenmeyer flasks (~897 mL), and then molecular grade agarose is added based on volume to final 0.8 %.

6. For autoclaving: heavy foil 10-in. square is folded over twice to make a foil cap for the two Erlenmeyer flasks, attached by autoclave tape.
7. The plate preparation agarose and top agarose are autoclaved on liquid 20 min sterilization cycle, then plate agarose is cooled to 60 °C before pouring into phytatrays 50 mL per tray.
8. Plates, after pouring, immediately have lids placed on loosely, and are allowed to set in clean hood 6 h to overnight. Then, lids are placed on fully/firmly and put upside-down inside sterile Tupperware storage box in 4 °C.
9. Top agarose is bench-cooled, swirled occasionally (always nitrile gloves on) as it is cooling down, then lid tightened and placed at 4 °C for storage.
10. Plate life is 3 weeks (to avoid fungus issues). Top agarose life is four uses (melt then cooling) within 8 weeks. All materials are stored at 4 °C when not directly in use.

3.2 Planting Seed

1. Mature dry seeds are premeasured (Fig. 2b) in one of two ways: 100 μL is enough to substantially cover the bottom of a phytatray yet not be so dense it may shield UV; or in alternative, if rows or disks of plants are needed, 30–50 seeds are individually counted out on wax paper squares.
2. Black boxes are wiped out with sterile wipe to remove dust or any errant spores (Fig. 2c).
3. Nitrile gloves wiped with 70 % ethanol are to be on henceforth for all procedures.
4. Top agarose (0.5× MS/MES pH 5.8; 0.8 % low melt agarose) is heated in microwave with rotation until melted fully. Plates are set out to be pre-acclimated to room temperature (Fig. 2d, e). Any excess condensation is shaken off from plates and lids.
5. Melted top agarose is then placed in 50 °C bath in the dark growth room (Fig. 2f).
6. Premeasured seeds are bleach sterilized (60 % bleach, 0.04 % triton X-100, autoclaved water) for 5 min at a ratio of 2.5 mL/100 μL seed, mixed by inversion.
7. Seeds are washed with sterile autoclaved water three times (inversion with parafilm to cover, settle seeds, pour or aspirate off with sterile pipette).
8. At last removal of water, where seeds are in ~0.5 mL of wash, 5 mL of top agarose (maintained in the 50 °C water bath) is added, mixed by inversion then immediately poured (or pipetted with a cut tip to reduce sheer force for small seed amounts) out onto the 0.5× MS plate (*see* **Note 4**).

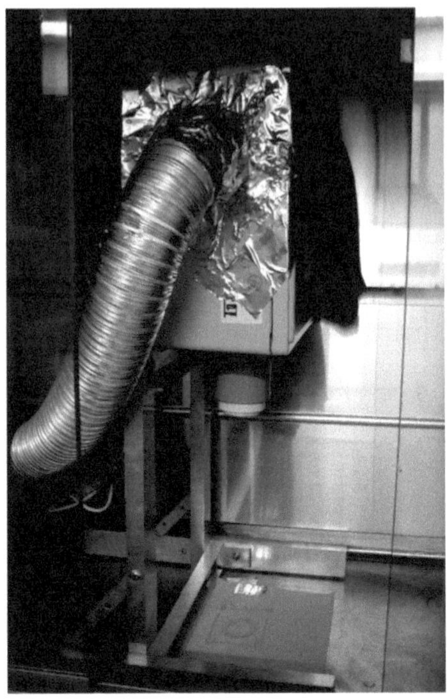

Fig. 3 The UV source for irradiations, with area marked for placement of phytatray

9. Plate is orbitally rotated gently by hand on the benchtop to spread seeds evenly across the plate, then lid placed on plate (Fig. 2h).
10. If disks or rows are desired instead of whole plate planting, measured seeds are in top agarose at same dilution degree as for full plate, applied by micropipet (1 mm cut from the dispo tip).
11. Poured seeds need to set for 5 min. Then plates are set into a black box (Fig. 2g–i).
12. The box is sealed then covered with foil and that foil taped to the box, then placed in 4 °C (Fig. 2j).
13. 48 h later, box is transferred unopened to 20 °C for 6–7 days (Arabidopsis; Soybean is not given the cold treatment and is maintained for 8 days at 20 °C).

3.3 Irradiations

1. Box is retrieved from −20 °C dark room/dark incubator on day 6 and brought into darkroom where light source is kept (also −20 °C dark room) for irradiation.
2. Lid is removed from plate and placed under pre-warmed-up (15 min, output checked) UV source for irradiation (Fig. 3; irradiation spot indicated).
3. For a low fluence (dose) treatment, irradiation in the range of 10^0 to 10^4 μmol/m^2 [1, 9–11] is given to plants.

4. The irradiation period is typically between 4 and 18 min depending on wavelength (fluence rate of source must be known to calculate).
5. Control plates are given a mock irradiation where lid is removed and plate is put in the irradiation position for the same period of time as the irradiated plate.
6. Plates are returned to the black box, box is resealed and wrapped and returned to growth room (*see* **Note 5**).

3.4 Experiments

1. Natural Fluorescence by deconvolution microscopy (in dark room with dim green safelight).
   (a) Experiments can be done or assessments made anywhere from 0 min after irradiation up to 24 h.
   (b) Entire live seedlings: whole live plants or sections can be placed on ethanol pre-cleaned glass slide.
   (c) Seedlings must be at least 1 mm apart, with a drop of sterile water.
   (d) Excess water is wicked away completely by microscope paper and periphery of slide is dried by microscope paper.
   (e) Clear nail polish is placed on dry slide, one layer around edge to accommodate coverslip.
   (f) Let dry until sticky (20–30 s).
   (g) Place 22 × 40 mm No. 1.5 ethanol-cleaned coverslip on seedlings.
   (h) Press down by first and second knuckle of index finger, rolling laterally to press down evenly.
   (i) Set in box, cover with foil to keep all light away until microscopy excitation.
   (j) Microscopy must be commenced within minutes after mounting.
   (k) Pre-warmed up and set up Deconvoluting arrangements should include DAPI-LongPass filter (i.e., must have UV bulb) for real color analysis; for apotomized sections false color is used with DAPI, FITC and Texas Red filter sets, where photography is software controlled (e.g., Zeiss Axiovision).
   (l) 10× visualizes entire cotyledon and partial stem of 7 day seedling; 20× only the cotyledon.
   (m) Optical sectioning or snaps can be done on mounted living seedlings.
   (n) Natural and False color fluorescence changes compared to wt are important phenotypic indicators for cotyledon and root tip.

2. Assessment of whole plant responses to UV.
   (a) To avoid light exposure that may cause responses, for any whole plant assessment, all equipment must be prepared prior to light exposure where photography or photocopying will take place.
   (b) All labels should be preprepared to be placed in a photograph or on photocopier, so identity, date and time of replicate or treatment is recorded at the time.
   (c) For statistical purposes 30 plants should be surveyed for phenotype for each treatment condition per replicate, with three to four biological replicates recommended.
   (d) At 0–24 h post-treatment, seedlings are best photographed on a stereomicroscope/dissecting scope in balanced white lighting, or imaged for details of morphology, shoot and root lengths and other characteristics.
3. RNA extraction for analysis involving RNA (in dark room with dim green safelight).
   (a) For planting, 100 μL of seed are planted per tray (*see* Subheading 3.2).
   (b) Each tray must be a separate treatment (e.g., Untreated, UV 300 nm, etc.)
   (c) Treatment and handling of plants is the same as described in Subheadings 3.2 and 3.3 (Fig. 1).
   (d) By this planting method only expression in aerial portions can be assessed.
   (e) On day 6, UV treatment is administered, plants returned to black box until harvest.
   (f) Prepare autoclaved foil (gloves on) by labeling the experiment samples on the foil with Sharpie. Place all foil squares into a larger piece of autoclaved foil to save until harvest.
   (g) At time 0 up to 24 h later, plants are removed from black box in the dark room for harvest of tissue.
   (h) Premeasured out liquid nitrogen (lN_2) in a lN_2 bucket is brought into the dark room.
   (i) lN_2 is poured into the plates enough to completely cover seedlings.
   (j) While lN_2 is evaporating off, an autoclaved foil square is fitted around the kitchen spatula to completely cover the end like a glove for the spatula.
   (k) Dip the spatula in the lN_2 for 10 s in order to get it very cold (seedlings won't stick).

(l) As soon as the last of the lN_2 is evaporated off, scrape pressing the bottom of the phytatray where the seedlings, like glass, will break off at the base of the stem (roots left in top agarose), from one end of the tray to the other.

(m) Add a small amount of lN_2 into the tray, which will keep all frozen.

(n) Tissue can be then poured into a prefrozen (with lN_2) mortar with pestle for grinding to powder in lN_2 or placed into the aluminum precut squares (**step f**).

(o) Whether ground in mortar and pestle or in foil packets, material must be kept frozen and must be kept in lN_2 until placed into –80 °C for storage.

(p) Ground material can also be extracted directly by methods described in [12, 13] or other RNA extraction recipe of choice.

(q) RNA is suitable for cDNA synthesis and analysis or direct assessment of RNA species.

4. Enzymatic activity and biochemical assessment.

(a) On day 6 or later in the dark room (time 0 or later from treatment) aerial portions of seedlings are cut with sterile scissors harvested directly into ice-cold buffer (dependent n what assay is to be done; [7, 12, 14–16]) with all preparatory steps in the dark.

(b) For a full tray of seedlings (100 μL original seed) 2.0 mL of buffer is required (1.5 mL extraction, 0.5 mL reextraction, and rinse of homogenizer).

(c) Protease inhibitors specific for plants and PMSF are needed for extended assays of protein activity [14, 15].

(d) Tissues are ground in a pointed-end ice-cold glass tissue homogenizer, with twisting and plunging to ensure all primary leaves ground.

(e) Homogenate is spun at 4 °C at 2,000 × g for 5 min to remove debris.

(f) At this point, with this concentrated homogenate other methods of extraction can be used for pigments (other solvents), etc.

(g) For plasma membrane assays (minus organelles; for a membrane fraction) an additional spin in a benchtop ultracentrifuge for small volumes is required by two-phase aqueous partitioning [16]. Percoll gradients can be used for investigation of specific organelles.

4 Notes

1. By these methods (Fig. 1) of studying UV signaling, we have grown the following plants in this manner to obtain useful data: *Arabidopsis*, *Pisum*, *Glycine max*, *Oryza*, *Triticum*, alfalfa, *Hordeum*, cotton, *Zea mays*, *Avena*.
2. Critical to truly examining UV effects is the quality of dark room and quality of UV source (Subheadings 2.1 and 2.2). "Light-tight" growth rooms and treatment areas were used for planting and maintaining seed, where the entire experiment from day 0 to day 7 (for Arabidopsis) is carried out in complete darkness, aided at times by a very dim green safelight.
3. In order to look directly at UV responses, minimal media is required where no salts, hormones, sugars (i.e., no sucrose), or other chemicals are added to the 0.5× MS/MES pH 5.8 plates (Subheadings 2.4 and 3).
4. When planting seed (Subheading 3.2) of Arabidopsis or other small seed (i.e., excluding soybean), rows are oriented perpendicular (across) to the short side of the plate.
5. At no time is the box or plate illuminated in any way even as planted seeds, but only brief periods in dim green safelight in the dark room. All preparations of materials and experiments must take place in darkness (until reading of Spec or on microscope for example) and for extracts, at 4 °C (Subheading 3).

References

1. Mandoli DF, Briggs WR (1981) Phytochrome control of two low-irradiance responses in etiolated oat seedlings. Plant Physiol 67:733–739
2. Lipeniec L, Debeaujon I, Routaboul JM, Baudry A, Pourcel L, Nesi N, Caboche M (2006) Genetics and biochemistry of seed flavonoids. Annu Rev Plant Biol 57:405–430
3. Searles PS, Flint SD, Caldwell MM (2001) A meta-analysis of plant field studies simulating stratospheric ozone depletion. Oecologia 127: 1–10
4. Frohnmeyer H, Staiger D (2003) Ultraviolet-B radiation-mediated responses in plants. Balancing damage and protection. Plant Physiol 133:1420–1428
5. Ulm R, Nagy F (2005) Signaling and gene regulation in response to ultraviolet light. Curr Opin Plant Biol 8:477–482
6. Sullivan JH, Gitz DC III, Liu-Gitz L, Xu C, Gao W, Slusser J (2007) Coupling short-term changes in ambient UV-B levels with induction of UV-screening compounds. Photochem Photobiol 83:863–870
7. Warpeha KM, Gibbons J, Carol A, Slusser J, Tree R, Durham W, Kaufman LS (2008) Presence of adequate phenylalanine mediated by G-protein is critical for protection from UV radiation damage in young etiolated Arabidopsis thaliana seedlings. Plant Cell Environ 31:1756–1770
8. Alonso JM, Stepanova AN, Leisse TJ et al (2003) Genome-wide insertional mutagenesis of Arabidopsis thaliana. Science 301:653–657
9. Briggs WR, Mandoli DF, Shinkle JR, Kaufman LS, Watson JC, Thompson WF (1984) Phytochrome regulation of plant development at the whole plant, physiological, and molecular levels. In: Columbetti G, Lenci F, Song P-S (eds) Sensory perception and transduction in aneural organisms, vol 89, NATO ASI Series. Series A: Life sciences. Plenum, New York. ISBN 89
10. Warpeha KMF, Kaufman LS (1990) Two distinct blue-light responses regulate epicotyl elongation in pea. Plant Physiol 92:495–499
11. Warpeha KMF, Kaufman LS (1990) Two distinct blue-light responses regulate the levels of

transcripts of specific nuclear-coded genes in pea. Planta 182:553–558

12. Warpeha KM, Upadhyay S, Yeh J, Adamiak J, Hawkins SI, Lapik YR, Anderson MB, Kaufman LS (2007) The GCR1, GPA1, PRN1, NF-Y signal chain mediates both blue light and ABA responses in Arabidopsis. Plant Physiol 143: 1590–1600
13. Thompson WF, Everett M, Polans NO, Jorgensen RA, Palmer JD (1983) Phytochrome control of RNA levels in developing pea and mung bean leaves. Planta 158:487–500
14. Warpeha KW, Lateef SS, Lapik YL, Anderson MB, Lee B-S, Kaufman LS (2006) G-protein-coupled Receptor1, G-protein Ga-Subunit1, and prephenate dehydratase1 are required for blue light-induced production of phenylalanine in etiolated Arabidopsis. Plant Physiol 140: 844–855
15. Warpeha KMF, Hamm HE, Rasenick MM, Kaufman LS (1991) A blue-light activated GTP-binding protein in plasma membrane of etiolated pea. Proc Natl Acad Sci U S A 88: 8925–8929
16. Warpeha KMF, Kaufman LS, Briggs WR (1992) A flavoprotein may mediate the blue light-activated binding of guanosine 5′-triphosphate to isolated plasma membranes of *Pisum sativum* L. Photochem Photobiol 55: 595–603

Chapter 11

Functional Analysis of Small Rab GTPases in Cytokinesis in *Arabidopsis thaliana*

Xingyun Qi and Huanquan Zheng

Abstract

Rab proteins are key regulators of membrane transport in eukaryotes. Recent evidence from different species supports the notion that some Rab proteins are crucial for cytokinesis, a pivotal procedure for successful cell division. As a family of monomeric small GTPases of the Ras superfamily, the function of Rab proteins is modulated by guanine nucleotide binding and hydrolysis. To investigate the function of Rab proteins, creating dominant negative or constitutively active mutant forms of a Rab protein is a widely used approach. To study cytokinesis in plant cells, using fluorescent dye to highlight the cell shape and the nuclei, and to monitor the formation of the newly formed cell plate in mitotic cells, is easy and useful. In this chapter, we describe detailed methods for (1) generating transgenic plants expressing dominant negative or constitutively active form of RAB-A1c; (2) fluorescent staining of cell shape, cell wall, and nuclei of mitotic root tip cells; (3) fluorescent staining of newly formed cell plate; and (4) detecting fluorescent signals using Confocal Laser Scanning Microscopy in the genetic model plant species *Arabidopsis thaliana*.

Key words Rab GTPases, Cytokinesis, Propidium iodide (PI), Calcofluor, FM4-64, *Arabidopsis thaliana*

1 Introduction

Rab proteins are key regulators of vesicular transport. They act as a molecular switch in regulating tethering/docking of different vesicles to a specific target membrane by cycling between their GTP-bound active and GDP-bound inactive forms [1–3]. As the largest family of monomeric small GTPases of the Ras superfamily, Rab proteins contain a highly conserved GTPase module, which binds either GTP or GDP nucleotides [4, 5]. Several conserved sequence elements in this module play key roles in regulating the binding of guanine nucleotides and the hydrolysis of bound GTP to GDP. For example, the phosphate/Mg^{2+} binding 1 (P/M1) element has a consensus sequence of GxxxxGKS/T, which interacts with the charged α- and β-phosphate groups of bound guanosine polyphosphate. A substitution of the conserved serine/threonine residue

Mark P. Running (ed.), *G Protein-Coupled Receptor Signaling in Plants: Methods and Protocols*, Methods in Molecular Biology, vol. 1043, DOI 10.1007/978-1-62703-532-3_11, © Springer Science+Business Media, LLC 2013

with asparagine produces a deactivated Rab mutant (SN mutant), with the protein locked into the GDP-bound state [6–10]. The P/M2 element is a highly conserved threonine necessary for Mg^{2+} coordination to the β- and γ-phosphate groups of GTP. The P/M3 element has a consensus sequence of WDTAGQ, which is involved in stabilizing a H_2O molecule coordinated to Mg^{2+}. A substitution of glutamine in this motif with leucine makes the motif unable to activate water as a nucleophile; therefore, the mutated protein is unable to hydrolyze GTP, rendering it an activated mutant (QL mutant) in the GTP-bound state [9–11]. It is believed that the GDP-locked SN mutant of a Rab protein potentially affects its upstream guanine nucleotide exchange factor [12], whereas the GTP-locked QL mutant of a Rab protein affects the downstream effectors [13]. Both the SN and QL mutated forms of Rab proteins are routinely used to investigate the function of the respective wild type Rab proteins [14–16].

Compared to the number of Rab proteins in *Saccharomyces cerevisiae*, Rab proteins in higher plants are greatly expanded. In the genome of *Arabidopsis*, 57 Rab loci were identified [17]. Recent evidence indicated that, similar to the function of some Rab GTPases in animal cells [18, 19], some plant Rab GTPases, as well as the putative upstream regulatory factors are involved in cytokinesis in *Arabidopsis* [14, 20, 21]. Cytokinesis is a process that separates the genetic material and the cytoplasmic organelles into the two daughter cells. In plant cells, cytokinesis requires the formation of a structure called the cell plate in the center of the phragmoplast [22–24]. During cytokinesis, numerous secretory vesicles from Golgi/TGN are delivered to the phragmoplast and fuse with each other to generate a tubular network, which is then morphologically restructured into a disk-like membrane compartment called the cell plate [23, 25]. Continual flow of vesicles to the margin of the newly formed cell plate allows the expansion of the cell plate, which eventually fuses with the preexisting plasma membrane [26].

In *Arabidopsis*, study of the function of a gene or a protein can be carried out in T-DNA insertional and/or RNAi mutants. Here, we describe in detail a pOp6/LhGR-based inducible expression of GDP- and GTP-locked forms of the RAB-A1c protein in plants to study the function of RAB-A1c, a member of the Rab-A1 subclass in cytokinesis. pOp6 is an artificial promoter that can be activated by LhGR, a synthetic transcription factor generated by fusing the ligand-binding domain of a rat glucocorticoid receptor to the amino terminus of the synthetic transcription factor LhG4 [27], thus LhGR can only be fully activated by dexamethasone [27].

The formation of the cell plate during cytokinesis in plants can be monitored by proteins involved in the process, for example, proteins such as KNOLLE (a syntaxin protein), RAB-A1c and RAB-A2a have been used to examine the cell plate formation [14, 20, 21, 28–30]. However, to use these proteins to study cytokinesis, one needs to either express fluorescence tagged version of these proteins, or generate specific antibody for the detection of these marker proteins in the mutant of interest. A much easier and faster way is to do a fluorescence based staining of the cell wall, plasma membrane and nuclei to monitor cytokinesis and the formation of the cell plate in the phragmoplast. Propidium iodide (PI), a red fluorescent dye normally used to stain the nucleic acid in animal cells, is typically used in plants to stain the cell wall of healthy cells, thus cell morphology and viability can be examined [20]. However, to visualize cell shape and the nuclei at the same time, a green fluorescent dye called calcofluor could be used to stain the cellulose in the cell wall and PI could be used to stain the chromosomal DNAs [21]. Finally, dynamics of the growing cell plate can be monitored by FM4-64, a dye that is inserted into the plasma membrane, and subsequently internalized to the *trans*-Golgi network/early endosome, where part of the dye will be targeted to the newly formed cell plate [14, 20, 21, 30, 31]. In this chapter, we describe in detail how to use PI to visualize cell morphology and viability; how to use calcofluor and PI to analyze cell division; and how to use FM4-64 to visualize the growing cell plate. These techniques can be used for the functional characterization of not only a Rab protein, but also perhaps any upstream regulators and downstream effectors [20, 29].

2 Materials

2.1 Materials

1. *Arabidopsis thaliana* seeds.
2. Disposable 1.5 mL Eppendorf tubes.
3. Disposable 15 mL plastic tubes.
4. Cloning vector pBluescript KS.
5. QuickChange™ Site-Directed Mutagenesis Kit.
6. pOp6 inducible expression vector PV-TOP vector [27].
7. LhGR driver line 4C-S5/7 (kanamycin resistant) [27].
8. AT (*Arabidopsis thaliana*) plates [32].
9. 100 mg/mL hygromycin solution.
10. 30 mg/mL kanamycin solution.
    (a) Weigh 0.3 g kanamycin powder and transfer it into a 15 mL plastic tube.
    (b) Add 10 mL sterilized distilled water (DDW) into the tube, stir well.
    (c) Sterilize the solution by passing the solution through a 0.45 μm sterilizing filter.
    (d) Keep the solution at −20 °C.
    (e) Melt completely before use.

11. 100 mmol/L dexamethasone solution.
    (a) Weigh 0.039 g dexamethasone powder into a 1.5 mL plastic tube.
    (b) Add 1 mL DMSO into the tube, stir well.
    (c) Keep the solution at −20 °C. Protect from light.
    (d) Melt completely before use.
12. Soil (Sunshine#5).
13. 1 mg/mL propidium iodide (PI) solution.
14. 3.5 mg/mL calcofluor solution ($C_{40}H_{44}N_{12}O_{10}S_2$).
    (a) Weigh 3.5 mg calcofluor powder into a 1.5 mL plastic tube.
    (b) Add 1 mL DDW in the tube.
    (c) Keep the solution at −20 °C. protect from light.
    (d) Melt completely and stir well before use.
15. 1 mol/L $CaCl_2$ solution.
16. Triton X-100.
17. 1 mol/L Tris buffer, pH 8.
18. 10 mg/mL RNase.
19. 5 mmol/L FM4-64 solution (*N*-(3-triethylammoniumpropyl)-4-(6-(4-(diethylamino)-phenyl) hexatrienyl) pyridinium dibromide).
    (a) Add 32.92 μL DDW to 100 μg FM4-64 powder provided by Molecular Probe in a tube to make a 5 mmol/L FM4-64 solution.
    (b) Store the solution at −20 °C, protect from light.
20. Microscope slides (1.0 mm thick) and microscope cover slips (0.13–0.17 mm thick).

2.2 Equipment

An inverted Zeiss LSM 510(meta) confocal laser scanning microscope (Zeiss, http://www.zeiss.com) equipped with 405 nm Diode laser, Argon (458, 477, 488, 514 nm), 543 nm HeNe laser and 633 nm HeNe laser.

3 Methods

3.1 Generation of Transgenic Plants Expressing Dominant Negative or Constitutively Active Mutants of RAB-A1c

1. Collect 7-day old wild type seedlings to extract total RNAs and do a RT-PCR using the primers listed below (restriction enzyme sites are underlined or in bold) to amplify the cDNA of *RAB-A1c*.

RAB-A1c-cDNA forward	GGGAATTC**GTCGAC**ATG GCGGGTTACAGAGC
RAB-A1c-cDNA reverse	GCGGATCC**GAGCTC**TTA GTTCGAGCAGCATCC

2. Clone the cDNA into the cloning vector pBluescript KS using EcoRI and BamHI, and sequence it.
3. Design overlapped PCR primers that introduce the desired mutations in *RAB-A1c* (listed below, mutated nucleotides are in bold, changed amino acid codons are underlined). For instance, to make the GDP-locked SN mutant of RAB-A1c, simply design a pair of primers that will encode the element GDSGVGK**N** instead of GDSGVGK**S**. To make the GTP-locked QL mutant of RAB-A1c, simply design a pair of primers that will encode the element WDTAG**L** instead of WDTAG**Q** (*see* **Note 1**).

RAB-A1c-SN forward	CAGGTGTGGGCAAAAAC AATTTGCTTTCACG
RAB-A1c-SN reverse	CGTGAAAGCAAATTGTT TTTGCCCACACCTG
RAB-A1c-QL forward	GGGATACTGCTGGTCTA GAAAGGTACCGAGCC
RAB-A1c-QL reverse	GGCTCGGTACCTTTCTA GACCAGCAGTATCCC

4. Do a temperature-cycling reaction using the primers listed in **step 3** and the cloned wild type *RAB-A1c* cDNA in pBluescript KS as the template; transform the amplified DNA into *E. coli* using QuickChange™ Site-Directed Mutagenesis Kit.
5. Isolate plasmid DNA from selected *E. coli* colonies and sequence to confirm modified SN or QL mutants of *RAB-A1c*.
6. Subclone the confirmed SN or QL mutant forms of *RAB-A1c* into the pOp6 inducible expression vector PV-TOP using SalI and SacI.
7. Transform the expression vector PV-TOP containing SN or QL mutant forms of *RAB-A1c* into Agrobacterium.
8. Transform the SN or QL mutant forms of *RAB-A1c* into the LhGR driver line 4C-S5/7 by flower dip [33]. Harvest seeds of the transformed plants.
9. Plate the harvested seeds in **step 8** on AT plates containing 15 μg/mL hygromycin (for pOp6-driven transgene) and 30 μg/mL kanamycin (for LhGR) to screen transgenic plants expressing the transgene.

10. Select approximately 20 individual transgenic plants resistant to hygromycin and kanamycin, and transfer them into soil (Sunshine#5) (*see* **Note 2**).
11. Grow plants at 22–24 °C under continuous light (80–100 μE/m s photosynthetically active radiation), and harvest T1 seeds.
12. Plate approximately 100 T1 seeds on AT plates containing 15 μg/mL hygromycin (for transgene), 30 μg/mL kanamycin (for LhGR) and 20 μmol/L dexamethasome for the segregation of the transgene and analysis of developmental phenotypes (e.g., primary root growth). At least two independent lines need to be selected for the experiments described below.

3.2 PI Staining (To Visualize Cell Morphology and Viability)

1. Plate *Arabidopsis* seeds on AT media, incubate vertically in the growth chamber at 22–24 °C under continuous light (80–100 μE/m s photosynthetically active radiation) for 5 days (*see* **Note 3**).
2. Take one seedling from the AT media and place it on a clean microscope slide.
3. Immediately pipet one drop of PI (1 mg/mL) on the seedling on the slide and incubate for 1 min at room temperature (*see* **Note 4**).
4. Cover the seedling with a clean cover slip.
5. The seedling is now ready for confocal laser scanning microscopy.

3.3 Calcofluor and PI Staining (To Analyze Cell Division)

1. Grow seedlings as described in **step 1** of Subheading 3.2.
2. Freshly prepare 1 mL staining solution as below (*see* **Note 5**).

   In a 1.5 mL Eppendorf tube, add 728 μL DDW, 10 μL calcofluor solution (3.5 mg/mL, stir well before use), 200 μL $CaCl_2$ (1 mol/L), 10 μL PI (1 mg/mL), 1 μL Triton X-100 and 50 μL Tris buffer (1 mol/L, pH 8). Finally add 1 μL RNase (10 mg/mL) (*see* **Note 6**).
3. Take one seedling from the AT media and immerse it in the freshly prepared staining solution.
4. Incubate the seedling for 60 min at room temperature.
5. Take the seedling from the staining solution and mount it in freshly prepared staining solution on a clean microscope slide. Cover it with a clean cover slip.
6. The seedling is now ready for confocal laser scanning microscopy.

3.4 FM4-64 Staining (To Visualize the Growing Cell Plate)

1. Grow seedlings on AT plates as described in **step 1** in Subheading 3.2.
2. Freshly prepare 1 mL 5 μmol/L FM4-64 solution as below.

In a 1.5 mL Eppendorf tube, add 999 μL DDW and 1 μL FM4-64 solution (5 mmol/L). Stir well.

3. Take one seedling from the AT media and immerse it in the 5 μmol/L FM4-64 solution for 30 min (*see* **Note 7**).
4. Take the seedling stained in the FM4-64 solution onto a clean microscope slide.
5. Pipet one drop of DDW to mount the seedling on the slide. Cover it with a clean cover slip.
6. The seedling is now ready for confocal laser scanning microscopy.

3.5 Confocal Laser Scanning Microscopy

1. Turn on the inverted Zeiss LSM 510(meta) confocal laser scanning microscope 10 min before use.
2. Open the Zeiss LSM510 software and choose the <Expert mode>.
3. Turn on laser channels that will be used. For PI staining and FM4-64 staining, choose the 543 nm HeNe laser. For calcofluor staining, choose the 405 nm Diode laser.
4. Choose the 40×/1.3 oil or 63×/1.4 oil objective and adjust the condenser so the light fills the focal plane.
5. Open the track configuration window to set up track configuration: For PI staining or FM4-64 staining, use the 543 nm laser to excite the sample and a LP 560-nm emission filter to collect the signal. For calcofluor and PI staining, use both 405 and 543 nm lasers to excite the sample, a 420–480 nm emission filter to collect calcofluor signal and a LP 560 nm emission filter to collect the PI signal.
6. In the scan panel, choose a line-sequential mode for scanning.
7. Scan the sample with <Fast XY> (*see* **Notes 8** and **9**).
8. Image collection: use <Single> to capture a single still image; click <Z-stack> and select the start point <Mark First>, the end point <Mark Last> and the interval between planes to make optical sections over the sample for a 3D image; click <Time Series>, fill in the number of time points and select a time between sections to make a movie (*see* **Notes 10** and **11**).
9. The images collected could be exported as still pictures in either TIFF or JPEG format. The optical sections can be projected as a 3D, which can be saved as an AVI file, similar to movies collected.
10. The images and movies could be processed with the Zeiss LSM image browser, Image J and PHOTOSHOP CS2 for publication.

4 Notes

1. The overlapped PCR primers should anneal to the same sequence on opposite strands of the plasmid and contain the desired mutation in the middle of both primers. The overlapped primers should be 25–45 bases in length and there should be 15 bases of correct sequence on both sides.
2. At least 20 individual transgenic plants should be selected to cover a range of expression level of the transgene.
3. The AT plates should be placed vertically in the growth chamber so that the roots are growing on the surface of the medium. In this case, when seedlings are removed from the plate, the root tips remain intact.
4. To stain the cell wall by PI, high concentration (1 mg/mL) of PI solution works much better than low concentration of PI.
5. Make sure to stir the calcofluor solution before use as there may be precipitates at the bottom of the tube.
6. Make sure to add RNase to degrade RNAs in the cytosol. Otherwise there will be strong background due to the PI staining of RNAs in the cytosol.
7. FM4-64 is a dye that stains the plasma membrane first. In Arabidopsis root tip cells, within 5 min of staining, the dye is endocytosed into the early endosome/*trans*-Golgi network; in 30 min, growing cell plates in mitotic cells are labeled by the dye; after more than 60 min, the late endosomal compartment is marked by the dye. FM4-64 is thus used to stain different endomembrane organelles in Arabidopsis root tip cells according to the staining time.
8. When general cell morphology is examined with PI staining, 40×/1.3 oil or 63×/1.4 oil objective without <Zoom> is sufficient. However, when cell division and the growth of cell plate are examined with calcofluor/PI and FM4-64, respectively, due to the small size of root tip cells, 2–4 times zoom is required to study a possible defect in cytokinesis and the growth of cell plates.
9. For multiple-color images, before examining samples stained with two different dyes, check if there is any bleed-through signal in one channel from another using control samples stained with each single fluorescence dye.
10. For the nuclei and cell plate staining, it is better to collect a series of optical Z-stack sections and project the series of images into a 3D image for examination.
11. For functional analysis of a Rab protein, especially a Rab protein that may be involved in the growth of cell plate, it is often better to collect movies to examine the dynamics or growth rate of cell plate over a certain time in the mutant generated.

References

1. Cao X, Ballew N, Barlowe C (1998) Initial docking of ER-derived vesicles requires Uso1p and Ypt1p but is independent of SNARE proteins. EMBO J 17:2156–2165
2. Allan BB, Moyer BD, Balch WE (2000) Rab1 recruitment of p115 into a *cis*-SNARE complex: programming budding COPII vesicles for fusion. Science 289:444–448
3. Zerial M, McBride H (2001) Rab proteins as membrane organizers. Nat Rev Mol Cell Biol 2:107–117
4. Chardin P (1988) The ras superfamily proteins. Biochimie 70:865–868
5. Chardin P (1991) Small GTP-binding proteins of the ras family: a conserved functional mechanism? Cancer Cells 3:117–126
6. Farnsworth CL, Feig LA (1991) Dominant inhibitory mutations in the $Mg^{(2+)}$-binding site of RasH prevent its activation by GTP. Mol Cell Biol 11:4822–4829
7. John J, Rensland H, Schlichting I, Vetter I, Borasio GD, Goody RS, Wittinghofer A (1993) Kinetic and structural analysis of the $Mg^{(2+)}$-binding site of the guanine nucleotide-binding protein p21H-ras. J Biol Chem 268:923–929
8. Valencia A, Chardin P, Wittinghofer A, Sander C (1991) The ras protein family: evolutionary tree and role of conserved amino acids. Biochemistry 30:4637–4648
9. Kotzer AM, Brandizzi F, Neumann U, Paris N, Moore I, Hawes C (2004) AtRabF2b (Ara7) acts on the vacuolar trafficking pathway in tobacco leaf epidermal cells. J Cell Sci 117: 6377–6389
10. Lee GJ, Sohn EJ, Lee MH, Hwang I (2004) The *Arabidopsis* rab5 homologs rha1 and ara7 localize to the prevacuolar compartment. Plant Cell Physiol 45:1211–1220
11. Krengel U, Schlichting I, Scherer A, Schumann R, Frech M, John J, Kabsch W, Pai EF, Wittinghofer A (1990) Three-dimensional structures of H-ras p21 mutants: molecular basis for their inability to function as signal switch molecules. Cell 62:539–548
12. Chen X, Ernst SA, Williams JA (2003) Dominant negative Rab3D mutants reduce GTP-bound endogenous Rab3D in pancreatic acini. J Biol Chem 278:50053–50060
13. Ang AL, Folsch H, Koivisto UM, Pypaert M, Mellman I (2003) The Rab8 GTPase selectively regulates AP-1B-dependent basolateral transport in polarized Madin-Darby canine kidney cells. J Cell Biol 163:339–350
14. Chow CM, Neto H, Foucart C, Moore I (2008) Rab-A2 and Rab-A3 GTPases define a trans-golgi endosomal membrane domain in *Arabidopsis* that contributes substantially to the cell plate. Plant Cell 20:101–123
15. de Graaf BH, Cheung AY, Andreyeva T, Levasseur K, Kieliszewski M, Wu HM (2005) Rab11 GTPase-regulated membrane trafficking is crucial for tip-focused pollen tube growth in tobacco. Plant Cell 17:2564–2579
16. Olkkonen VM, Stenmark H (1997) Role of Rab GTPases in membrane traffic. Int Rev Cytol 176:1–85
17. Pereira-Leal JB, Seabra MC (2001) Evolution of the Rab family of small GTP-binding proteins. J Mol Biol 313:889–901
18. Sakamori R, Das S, Yu S, Feng S, Stypulkowski E, Guan Y, Douard V, Tang W, Ferraris RP, Harada A et al (2012) Cdc42 and Rab8a are critical for intestinal stem cell division, survival, and differentiation in mice. J Clin Invest 122: 1052–1065
19. Chesneau L, Dambournet D, Machicoane M, Kouranti I, Fukuda M, Goud B, Echard A (2012) An ARF6/Rab35 GTPase cascade for endocytic recycling and successful cytokinesis. Curr Biol 22:147–153
20. Qi X, Kaneda M, Chen J, Geitmann A, Zheng H (2011) A specific role for *Arabidopsis* TRAPPII in post-Golgi trafficking that is crucial for cytokinesis and cell polarity. Plant J 68: 234–248
21. Qi X, Zheng H (2013) Rab-A1c GTPase defines a population of *trans*-Golgi network that is sensitive to endosidin1 during cytokinesis in *Arabidopsis*. Mol Plant 6:847–859
22. Gu X, Verma DP (1996) Phragmoplastin, a dynamin-like protein associated with cell plate formation in plants. EMBO J 15:695–704
23. Samuels AL, Giddings TH Jr, Staehelin LA (1995) Cytokinesis in tobacco BY-2 and root tip cells: a new model of cell plate formation in higher plants. J Cell Biol 130:1345–1357
24. Segui-Simarro JM, Austin JR II, White EA, Staehelin LA (2004) Electron tomographic analysis of somatic cell plate formation in meristematic cells of *Arabidopsis* preserved by high-pressure freezing. Plant Cell 16: 836–856
25. Staehelin LA, Hepler PK (1996) Cytokinesis in higher plants. Cell 84:821–824
26. Mayer U, Jurgens G (2004) Cytokinesis: lines of division taking shape. Curr Opin Plant Biol 7:599–604
27. Craft J, Samalova M, Baroux C, Townley H, Martinez A, Jepson I, Tsiantis M, Moore I (2005) New pOp/LhG4 vectors for stringent glucocorticoid-dependent transgene expression in *Arabidopsis*. Plant J 41:899–918

28. Lauber MH, Waizenegger I, Steinmann T, Schwarz H, Mayer U, Hwang I, Lukowitz W, Jurgens G (1997) The *Arabidopsis* KNOLLE protein is a cytokinesis-specific syntaxin. J Cell Biol 139:1485–1493
29. Qi X, Zheng H (2011) *Arabidopsis* TRAPPII is functionally linked to Rab-A, but not Rab-D in polar protein trafficking in *trans*-Golgi network. Plant Signal Behav 6: 1679–1683
30. Boutte Y, Frescatada-Rosa M, Men S, Chow CM, Ebine K, Gustavsson A, Johansson L, Ueda T, Moore I, Jurgens G et al (2010) Endocytosis restricts *Arabidopsis* KNOLLE syntaxin to the cell division plane during late cytokinesis. EMBO J 29:546–558
31. Dhonukshe P, Baluska F, Hlavacka A, Schlicht M, Samaj J, Friml J, Gadella TWJ Jr (2006) Endocytosis of cell surface material mediates cell plate formation during plant cytokinesis. Dev Cell 10:137–150
32. Haughn GW, Somerville C (1986) Sulfonylurea-resistant mutants of *Arabidopsis thaliana*. Mol Gen Genet 204:430–434
33. Clough SJ, Bent AF (1998) Floral dip: a simplified method for Agrobacterium-mediated transformation of *Arabidopsis thaliana*. Plant J 16:735–743

Chapter 12

In Vivo Localization in Arabidopsis Protoplasts and Root Tissue

Myoung Hui Lee, Yongjik Lee, and Inhwan Hwang

Abstract

In eukaryotic cells, a large number of proteins are transported to their final destination after translation by a process called intracellular trafficking. Transient gene expression, either in plant protoplasts or in specific plant tissues, is a fast, flexible, and reproducible approach to study the cellular function of proteins, protein subcellular localizations, and protein–protein interactions. Here we describe the general method of protoplast isolation, polyethylene glycol-mediated protoplast transformation and immunostaining of protoplast or intact root tissues for studying the localization of protein in *Arabidopsis*.

Key words PEG-mediated transformation, Protoplast, Subcellular localization, Immunocytochemistry

1 Introduction

Vesicle trafficking is a finely controlled process that transports proteins and soluble molecules between endomembrane compartments [1]. Vesicle trafficking consists of diverse events, such as vesicle formation, motility, docking, and fusion, which are controlled by various protein factors [2–4]. Although the general principle of vesicle trafficking is highly conserved among all eukaryotes, the function of many key players involved in vesicle trafficking is still poorly understood in plants.

Generation of transgenic lines has become widely used to investigate the function of genes in plants [5, 6]. However, establishing a stable transgenic plant is still relatively expensive and time-consuming process, which limits the transgenic approach for the genome-wide functional analysis. Transient gene expression provides an alternative system for the functional analysis of plant genes [7]. Compared with the transgenic approach, transient expression system has a major advantage in its rapid and efficient analysis of gene functions. So far, various transient gene expression methods have been developed. Transient expression using virus infection or leaf infiltration with *Agrobacterium* is a simple and

Mark P. Running (ed.), *G Protein-Coupled Receptor Signaling in Plants: Methods and Protocols*, Methods in Molecular Biology, vol. 1043, DOI 10.1007/978-1-62703-532-3_12,

© Springer Science+Business Media, LLC 2013

effective method in the large-scale analyses of gene function in plants [8, 9]. Transient expression in protoplasts is another popular method to analyze gene expression, protein subcellular localization, protein–protein interaction, and protein activity [10–12]. In addition to its high transformation efficiency, protoplasts are able to respond to hormones and environmental cues in a similar manner as that in intact tissues, providing a powerful and versatile tool for the dissection of cellular processes [12].

Many cellular processes are spatially restricted within the cell. Specific trafficking pathways ensure that particular proteins are targeted to the right place in the cell [13, 14]. Therefore, identification of protein subcellular localization is important to understand the function of proteins involved in various cellular processes. Subcellular localizations of many proteins have not been characterized, and computational methods for predicting the subcellular localization of plant proteins are less accurate compared to those of prokaryotic proteins [15]. Immunolocalization has been used to determine the protein localization with high sensitivity and resolution [16, 17]. However, a specific antibody that is suitable for the immunohistological detection of a particular protein is not always readily available. Transient expression of a protein with a specific epitope tag in protoplasts can overcome this problem.

Here, we present methods regarding protoplast isolation, polyethylene glycol (PEG)-mediated transformation, immunocytochemistry of transformed protoplasts or intact root tissues, and laser scanning confocal microscopy.

2 Materials

2.1 Plants and Media

1. Arabidopsis (*Arabidopsis thaliana* ecotype Columbia) is grown on B5 medium plates (see below) at 22 °C in a growth chamber with a 16/8 h light/dark cycle. Leaf tissues obtained from 15- to 18-day-old plants are used for the protoplast isolation.
2. B5 medium plates: 3.2 g/L Gamborg's B5 medium (Duchefa Biochemie, Haarlem, The Netherlands), 2 % (w/v) sucrose, 0.05 % (w/v) 2-(*N*-morpholino)ethanesulfonic acid (MES) free-acid monohydrate, and 0.8 % (w/v) agar. The medium should be autoclaved for 15 min at 121 °C and then poured into Petri dishes.

2.2 Buffers and Reagents

1. Enzyme solution: 0.25 % (w/v) macerozyme R-10 (Yakult Honsha Co. Ltd., Tokyo, Japan), 1.0 % (w/v) cellulase R-10 (Yakult Honsha Co. Ltd.), 400 mM mannitol, 8 mM $CaCl_2$, and 5 mM MES-KOH, pH 5.6 (*see* **Note 1**).
2. 21 % (w/v) sucrose solution (*see* **Note 2**).

3. W5 solution: 154 mM NaCl, 125 mM $CaCl_2$, 5 mM KCl, 5 mM glucose, and 1.5 mM MES-KOH, pH 5.6 (*see* **Note 2**).
4. MaMg solution: 400 mM mannitol, 15 mM $MgCl_2$, and 5 mM MES-KOH, pH 5.6 (*see* **Note 2**).
5. PEG solution: 400 mM mannitol, 100 mM $Ca(NO_3)_2$, and 40 % (w/v) PEG 8,000 (*see* **Note 3**).
6. W6 buffer: 10 mM HEPES, pH 7.2, 154 mM NaCl, 125 mM $CaCl_2$, 2.5 mM maltose, and 5 mM KCl.
7. 16 % (w/v) paraformaldehyde solution.
8. TSW buffer: 10 mM Tris–HCl, pH 7.4, 0.9 % (w/v) NaCl, 0.25 % (w/v) gelatin, 0.02 % (w/v) sodium dodecyl sulfate (SDS), and 0.1 % (w/v) Triton X-100 (*see* **Notes 4** and **5**).
9. Mowiol 4-88 (Sigma-Aldrich, Saint Louis, MI, USA).
10. PEX buffer: 5 mM $MgCl_2$, 100 mM PIPES, and 5 mM EGTA, pH 6.9.
11. Fixing solution: PEX buffer, 4 % Paraformaldehyde (PFA), and 5 % DMSO.
12. Blocking solution: PEX buffer and 0.5 % BSA.
13. Driselase (Sigma-Aldrich, Germany).

2.3 Plasmid DNA

1. DNA constructs: clone the gene into a pUC-based plant expression vector.
2. Plasmid DNA should be purified using Qiagen Plasmid Maxi Prep columns (Qiagen, Cologne, Germany), or equivalent columns.

2.4 Equipment and Software

1. Mesh (140 μm of pore size). For example, cell dissociation sieve CD1 (Sigma-Aldrich, Saint Louis, MI, USA).
2. Poly-L-lysine-coated slide (Polysciences, Warrington, PA, USA).
3. PAP pen or materials of hydrophobic barrier.
4. Tabletop centrifuge and swinging-bucket rotor to accommodate 15 mL Falcon tubes. For example, Eppendorf 5702 (Eppendorf North America, Hauppauge, NY, USA). This is used throughout the protoplast isolation and transformation procedures.
5. Hemacytometer (Cole-Parmer, London, UK), and a microscope equipped for phase-contrast optics.
6. Zeiss LSM 510 META laser scanning confocal microscope (Jena, Germany) or equivalents.
7. Adobe Photoshop software (Mountain View, CA, USA).

3 Methods

All of the steps should be performed at room temperature, unless otherwise specified.

3.1 Isolation of Protoplasts from Arabidopsis Plants

1. Grow the plants for 15–18 days on a B5 plate.
2. Harvest leaf tissues (150 plants for ten transformation experiments) of Arabidopsis plants from the plate using a new scalpel.
3. Dip the leaf tissues into 20 mL of enzyme solution. At this step, the amount of enzyme solution should be barely enough to soak all the tissues.
4. Incubate for 8–12 h with gentle agitation (very slowly, at 23 °C, in the dark). After incubation, the solution should display a strong green color. If the incubation time is too long, the protoplasts will be stressed.
5. Filter the enzyme/protoplast solution through the 140 μm mesh to remove undigested leaf tissues.
6. Load the mixture on top of 20 mL of 21 % sucrose solution in a 50 mL Falcon tube.
7. Centrifuge at 730 rpm (98 × *g*) for 10 min in a swinging-bucket rotor. After centrifugation, the Falcon tube contains top and bottom fractions that contain enzyme solution and 21 % sucrose solution, respectively.
8. Transfer protoplasts in the top fraction and at the interface between top and bottom fractions into 30 mL of W5 solution in a 50 mL Falcon tube. At this step, it is preferable to use a Pasteur pipette or a blue tip (1 mL) with the tip (0.5 cm from the end) cut off by a scalpel. Be careful not to touch the bottom fraction because sucrose usually prevents intact protoplasts from being pelleted to the bottom at the next step.
9. Pellet the protoplasts by centrifugation at 530 rpm (51 × *g*) for 6 min in a swinging-bucket rotor.
10. Discard the supernatant completely.
11. Add 20 mL of W5 solution and suspend the protoplasts by gently inverting the tube.
12. Keep the protoplast at 4 °C for at least 1 h.

3.2 PEG-Mediated Transformation

1. Aliquot 20–30 μg of plasmid DNA into a 15-mL round-bottom tube.
2. Pellet the protoplasts by centrifugation at 500 rpm (46 × *g*) for 2–3 min in a swinging-bucket rotor.
3. Discard the W5 solution.

4. Suspend the protoplasts in ~3 mL MaMg solution at a density of 5×10^6 protoplasts/mL. The number of protoplasts is counted using a hemacytometer.
5. Add 300 μL of protoplasts to each 15 mL tube containing plasmid DNA.
6. Mix gently and completely by rotating the tubes several times at an almost horizontal position by hand. This step is critical because it will prevent protoplasts from being coagulated during next steps.
7. Add immediately 300 μL of PEG solution into the tube and mix well by rotating gently the tubes at an almost horizontal position by hand.
8. Incubate at room temperature (23 °C) for 30 min.
9. Add 0.6 mL of W5 solution, mix completely by rotating the tubes, and incubate for 10 min.
10. Add 1 mL of W5 solution, mix completely by rotating the tubes, and incubate for 10 min.
11. Add 2 mL of W5 solution, mix completely by rotating the tubes, and incubate for 10 min.
12. Add 4 mL of W5 solution and mix completely. The total volume will be 8.5 mL.
13. Centrifuge at 500 rpm ($46 \times g$) for 5 min in a swinging-bucket rotor. Discard the supernatant.
14. Add 2 mL of W5 solution and resuspend the protoplasts completely.
15. Incubate at 23 °C in an incubator.

3.3 Immunocytochemistry with Protoplasts

Subcellular localization of nonfluorescent reporter proteins can be determined by immunocytochemistry using antibody raised against proteins or epitopes.

1. Resuspend transformed protoplast in 300–900 μL W6 buffer (*see* **Note 6**).
2. Draw a circle on the poly-L-lysine-coated glass slides with PAP pen or materials of hydrophobic barrier.
3. Spread 300 μL of cell suspension within the circle of poly-L-lysine-coated glass slides and allow the protoplasts to adhere for 1 h at room temperature (*see* **Note 7**).
4. Fix the protoplasts with 4 % (v/v) paraformaldehyde and incubate for 1 h.
5. Permeabilize the protoplasts by washing three times with 300 μL of TSW buffer for 30 min (10 min for each wash) on the slide. No mixing or shaking is required during the washing step.

6. Incubate the protoplasts with primary antibody diluted in 300 μL of TSW buffer for 16 h at 4 °C. The dilution factor largely depends on the titer of the antibody used.
7. Wash the protoplasts three times with 300 μL of TSW buffer for 30 min (as described in **step 5**).
8. Incubate the protoplasts with the secondary antibody diluted in 300 μL of TSW buffer at room temperature for 3 h.
9. Wash the protoplasts three times with 300 μL of TSW buffer.
10. Mount with 70 μL of Mowiol 4-88 and cover with a coverslip (*see* **Note 8**). The coverslip can be sealed with clear nail polish or a commercial sealant to prevent sample damage.
11. Images are obtained by laser scanning confocal microscopy.

3.4 Laser Scanning Confocal Microscopy

1. Capture images with Zeiss LSM 510 META laser scanning confocal microscope (Jena) using C-APOCHROMAT (×40/1.2 W numerical aperture water immersion lens) in the multi-track mode or equivalent microscopes.
2. Excitation/emission wavelengths were 488/505–530 nm for FITC and GFP, and 543/560–615 nm for RFP and TRITC.
3. Capture transmitted light reference images using differential interference contrast optics and argon laser illumination at 488 nm.
4. Process images using Adobe Photoshop software for pseudo-color images.

3.5 Immunocytochemistry with Arabidopsis Seedlings

Immunofluorescence labeling of proteins in root cells is performed as described by Wee et al. [16] and Kircher et al. [17] with modification.

1. Place 3–5 day-old seedlings into a plate or plastic tube containing fixing solution and incubate for 60 min.
2. Wash the seedlings five times with 200 μL of PEX buffer.
3. Draw the outline on the poly-L-lysine-coated slide with PAP Pen or materials of hydrophobic barrier.
4. Place fixed seedlings onto poly-L-lysine-coated slides within the marked area and let them dry.
5. Digest the cell wall by treating the tissues with 200 μL of 2 % Driselase in PEX buffer for 40 min at 37 °C.
6. Wash the seedlings five times with 200 μL of PEX buffer.
7. For blocking nonspecific binding of antibody, incubate the seedlings with 200 μL of blocking buffer for overnight at 4 °C.
8. Wash the seedlings three times 200 μL of PEX buffer.
9. Incubate the seedlings with the primary antibody diluted in 200 μL of PEX buffer for 3 h at room temperature or overnight at 4 °C.

10. Wash the seedlings three times with 200 μL of PEX buffer.
11. Incubate the seedlings with the secondary antibody diluted in 200 μL of PEX buffer for 3 h at room temperature.
12. Wash the seedlings three times with 200 μL of PEX buffer.
13. Mount with 70 μL of Mowiol 4-88 and cover with a coverslip (*see* **Note 8**). The coverslip can be sealed with clear nail polish or a commercial sealant to prevent sample damage.
14. Examine the samples by laser scanning confocal microscopy, as described in Subheading 3.4.

4 Notes

1. Enzyme solution can be stored at −20 °C for a few months after filtration.
2. MaMg solution, W5 solution and 21 % sucrose are sterilized by autoclaving at 121 °C for 15 min and can be stored at room temperature, but it is recommended to store at 4 °C. These solutions are easily contaminated.
3. PEG solution can be stored at 37 °C for a week, but it is recommended to prepare freshly every time. Transformation efficiency is better with fresh PEG solution.
4. Detergents such as Triton X-100 are used to permeabilize the cell membrane for improving antibody penetration.
5. To reduce nonspecific staining in immunocytochemistry, the samples are incubated with a blocking buffer that contains normal serum, nonfat dry milk, or gelatin.
6. Suspension volume depends on the cell density.
7. The slide glass with samples should be stored in a foil-wrapped dish with water-soaked paper towel at the bottom.
8. After immunostaining, for long-term usage, the sample should be stored at 4 °C in a dark condition to prevent solubilization by enzymatic reactions or fluorophore photobleaching.

References

1. Juergens G (2004) Membrane trafficking in plants. Annu Rev Cell Dev Biol 20:481–504
2. Hwang I, Robinson DG (2009) Transport vesicle formation in plant cells. Curr Opin Plant Biol 12:660–669
3. Bröcker C, Engelbrecht-Vandré S, Ungermann C (2010) Multisubunit tethering complexes and their role in membrane fusion. Curr Biol 20:943–952
4. Nielsen E, Cheung AY, Ueda T (2008) The regulatory RAB and ARF GTPases for vesicular trafficking. Plant Physiol 147:1516–1526
5. Parinov S, Sundaresan V (2000) Functional genomics in Arabidopsis: large-scale insertional mutagenesis complements the genome sequencing project. Curr Opin Biotechnol 11:157–161
6. An G, Jeong DH, Jung KH, Lee S (2005) Reverse genetic approaches for functional genomics of rice. Plant Mol Biol 59:111–123
7. Denecke J, Aniento F, Frigerio L, Hawes C, Hwang I, Mathur J, Neuhaus JM, Robinson DG (2012) Secretory pathway research: the more experimental systems the better. Plant Cell 24:1316–1326

8. Tsuda K, Qi Y, le Nguyen V, Bethke G, Tsuda Y, Glazebrook J, Katagiri F (2012) An efficient Agrobacterium-mediated transient transformation of Arabidopsis. Plant J 69:713–719
9. Van Loock B, Markakis MN, Verbelen JP, Vissenberg K (2010) High-throughput transient transformation of Arabidopsis roots enables systematic colocalization analysis of GFP-tagged proteins. Plant Signal Behav 5: 261–263
10. Jung CJ, Lee MH, Min MK, Hwang I (2011) Localization and trafficking of an isoform of the AtPRA1 family to the Golgi apparatus depend on both N- and C-terminal sequence motifs. Traffic 12:185–200
11. Chen S, Tao L, Zeng L, Vega-Sanchez ME, Umemura K, Wang GL (2006) A highly efficient transient protoplast system for analyzing defence gene expression and protein-protein interactions in rice. Mol Plant Pathol 7: 417–427
12. Sheen J (2001) Signal transduction in maize and Arabidopsis mesophyll protoplasts. Plant Physiol 127:1466–1475
13. Reyes FC, Buono R, Otegui MS (2011) Plant endosomal trafficking pathways. Curr Opin Plant Biol 14:666–673
14. Hwang I (2008) Sorting and anterograde trafficking at the Golgi apparatus. Plant Physiol 148:673–683
15. Emanuelsson O, Brunak S, von Heijne G, Nielsen H (2007) Locating proteins in the cell using TargetP, SignalP and related tools. Nat Protoc 2:953–971
16. Wee EG, Sherrier DJ, Prime TA, Dupree P (1998) Targeting of active sialyltransferase to the plant Golgi apparatus. Plant Cell 10:1759–1768
17. Kircher S, Wellmer F, Nick P, Rügner A, Schäfer E, Harter K (1999) Nuclear import of the parsley bZIP transcription factor CPRF2 is regulated by phytochrome photoreceptors. J Cell Biol 144:201–211

Chapter 13

Analysis of Protein Prenylation and *S*-Acylation Using Gas Chromatography–Coupled Mass Spectrometry

Nadav Sorek, Amir Akerman, and Shaul Yalovsky

Abstract

Lipid modifications play a key role in protein targeting and function. The two Arabidopsis Gγ subunits, AGG1 and AGG2, have been shown to undergo prenylation (AGG1) and *S*-acylation (AGG2). Prenylation involves covalent nonreversible attachment of either farnesyl (15 carbons) or geranylgeranyl (20 carbons) isoprenoids to conserved cysteine residues at or near the C-terminus of proteins. *S*-acylation, frequently referred to as palmitoylation, involves the attachment of acyl fatty acids to thiol groups of cysteine residues through a reversible thioester bond. The procedures described below allow direct analysis of the prenyl and acyl moieties using gas chromatography–coupled mass spectrometry (GC-MS). These methods are based on (1) cleavage of prenyl groups with the Raney nickel catalyst and (2) analysis of protein *S*-acylation following cleavage of the acyl fatty acids from proteins by hydrogenation with platinum (IV) oxide. The hydrogenation under these conditions causes an acid transesterification of the acyl moieties, adding an ethyl group to the carboxyl head of the fatty acid. The addition of the ethyl group reduces the polarity of the fatty acids, allowing their efficient separation by gas chromatography.

Key words Prenylation, *S*-acylation, Palmitoylation, Gas chromatography–coupled mass spectrometry, Lipid analysis

1 Introduction

Lipid modifications play an important role in the subcellular targeting and function of heterotrimeric G proteins. Mammalian Gα subunits are either myristoylated or palmitoylated or modified by both lipids. Plasma membrane attachment of mammalian Gγ subunits depends on prenylation for plasma membrane localization and signaling [1]. It has been shown that in Arabidopsis AGG1 is prenylated, primarily by geranylgeranyl while AGG2 is likely *S*-acylated [2–5].

The methods described bellow allows chemical identification of both prenyl and acyl groups following their separation from purified recombinant proteins (Fig. 1). The analysis of protein prenylation is a modification of a previously described method [6].

Mark P. Running (ed.), *G Protein-Coupled Receptor Signaling in Plants: Methods and Protocols*, Methods in Molecular Biology, vol. 1043, DOI 10.1007/978-1-62703-532-3_13, © Springer Science+Business Media, LLC 2013

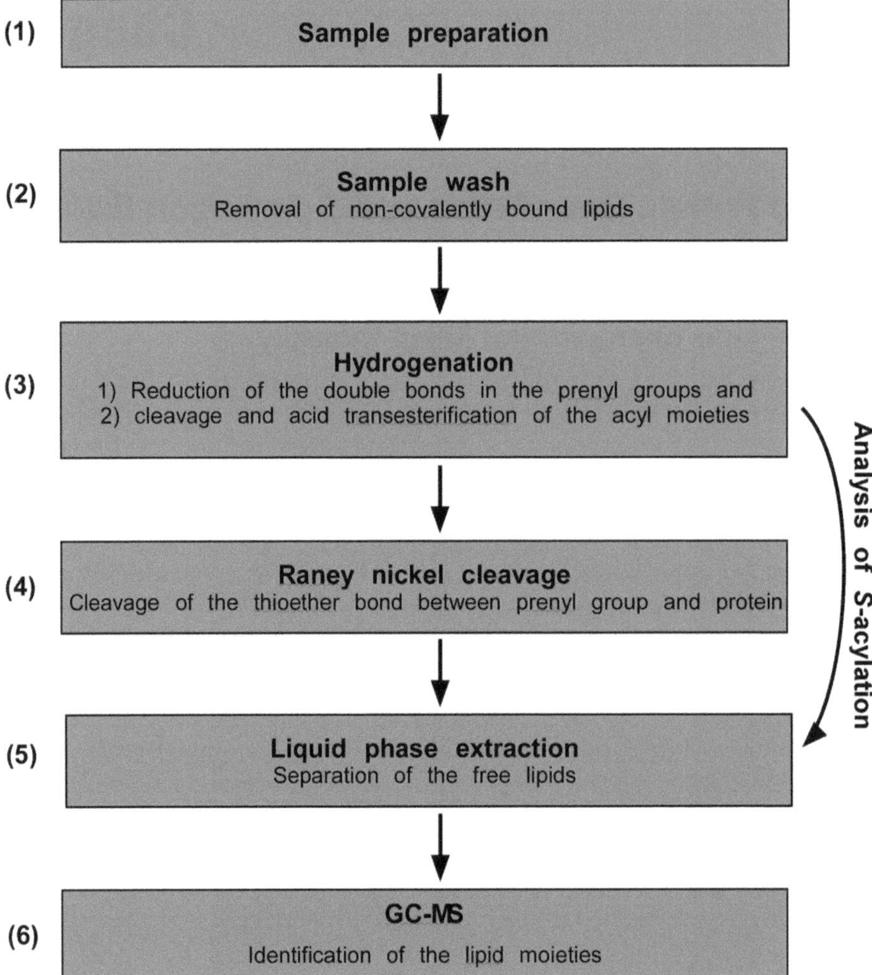

Fig. 1 An overview of the method. After protein purification, the sample is washed to remove any non-covalent attached lipids. Hydrogenation cause reduction of the prenyl double bonds but can also be used to cleave the acyl moieties form the protein. Raney nickel is then used to break the thioether bonds between the prenyl groups and the protein. Note that for analysis of *S*-acylation only, the Raney nickel cleavage is not required. The released lipid moieties are then extracted by a phase separation with an organic solvent such as pentane. The released and purified lipid moieties are in turn identified by GC-MS

This method is based on cleavage of the thioether bond between the prenyl groups and the proteins with Raney nickel [6, 7]. To enable identification of the released prenyl moieties, their double bonds are reduced using hydrogenation prior to Raney nickel cleavage (Fig. 2a) [6]. This step is necessary since the Raney nickel catalyst induces double bond shifts, leading to the formation of multiple compounds, which hamper identification by gas chromatography–coupled mass spectrometry (GC-MS). Following hydrogenation, the reduced released form of farnesyl will be detected as 2,6,10-trimethyldodecane and of geranylgeranyl as 2,6,10,14-tetramethylhexadecane [8]. The hydrogenation step also

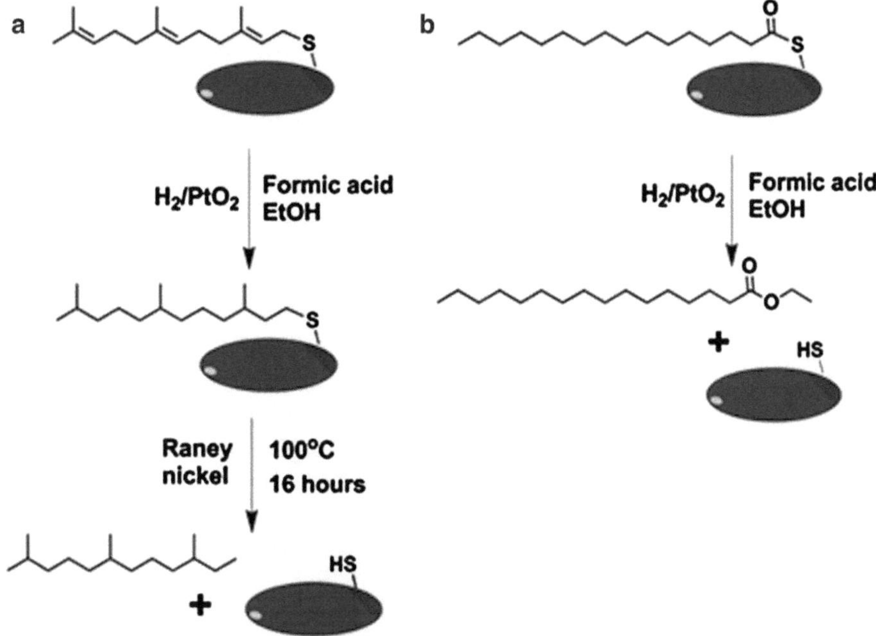

Fig. 2 The hydrogenation reaction mechanism and the Raney nickel cleavage. (**a**) The analysis of prenylation involved hydrogenation that causes reduction of the double bonds followed by Raney nickel cleavage, which break the thioether bond between the cysteine in the protein and the prenyl groups. Only farnesyl is illustrated but the procedure is the same for geranylgeranyl. (**b**) The acid transesterification reaction mechanism, which cleaves the acyl fatty acid from the protein concomitantly with the addition of ethyl group to the polar head. Only palmitic acid is illustrated but the same procedure can be used for any acyl moieties. In both (**a**) and (**b**) the *oval shape* corresponds to the modified protein

reduces thioester linkages between acyl groups and cysteines, facilitating their identification. The Raney nickel cleavage is not required for analysis of protein *S*-acylation [9].

The polar head group hinders identification of acyl fatty acids by GC-MS. The hydrogenation facilitates formation of ethyl modified fatty acid, overcoming the polarity barrier and therefore enables efficient identification of acyl lipids by GC-MS. To release the acyl lipids, proteins are dissolved in a solution of formic acid and ethanol and in turn the acyl thioester bond is cleaved by hydrogenation in the presence of platinum (IV) oxide [10], instead of typically used hydroxylamine treatments [11, 12]. The combination of acidic ethanol and hydrogenation treatments causes acid transesterification resulting in to ethyl esterification of the released acyl groups (Fig. 2b). In comparison, when fatty acids are released by hydroxylamine treatments, they are not efficiently separated by GC, forming wide peaks that are difficult to interpret [9].

The protocol consists of six major steps: (1) sample preparation, a purified protein should be used; (2) sample wash, to insure that there are no non covalent lipids bounding to the protein;

(3) hydrogenation, for reduction of the prenyl double bonds and/or acid transesterification of the acyl moieties; (4) Raney nickel cleavage, for cleaving the prenyl groups; (5) lipid extraction, to separate the lipids from the soluble proteins and (6) GC-MS analysis, to identify the exact lipid moieties.

As noted above, the Raney nickel cleavage is not necessary for the analysis of *S*-acylation and therefore if prenylation is not being analyzed, step 4 (Raney nickel cleavage) could be skipped and following step 3, the procedure should continue directly from step 5 (lipid extraction).

The protocol described here should be useful in the following instances: (1) following identification of putative prenylated or *S*-acylated proteins by high-throughput methods; (2) determining whether prenylation and/or S-acylation affect subcellular localization and targeting of proteins; (3) the identification and quantification of the modifying lipid groups of known prenylated or *S*-acylated proteins and (4) for analyzing changes in protein prenylation or *S*-acylation status following drug treatments, changes in growth conditions and comparison of proteins that were expressed in wild-type and mutant backgrounds. Follow-up experiments could include experiments such as: (1) testing how the lipid modifications affects the dynamics of interaction of protein with membranes, (2) analyzing whether the lipid modifications regulates protein function and to which activities is it required, (3) analyzing whether the lipid modifications are stable or transient and (4) attempting to identify the *S*-acyl transferases responsible for modifying a given protein.

Analysis of protein prenylation requires 25–100 μg of purified recombinant protein. Analysis of protein *S*-acylation can be carried out with 1–25 μg or more of purified recombinant protein. Importantly, the acyl groups can be easily identified also when using 1 μg of purified protein, while for prenylation analysis 25 μg is the minimum amount of protein required for analysis. Given the amounts of purified proteins that need to be obtained for the lipid identification, the method described here is not well suited for high-throughput analysis. The amount of starting material required for the analysis depends on the protein expression levels and therefore is subject to changes between different proteins. Relative quantification of the protein lipidation levels can be achieved provided that protein concentrations and volumes of extraction solutions are equal between different samples. Because recovery of the acyl and prenyl lipids from the GC column can vary, it is essential to compare all samples on the same day and repeat the measurements on different days. It is advisable to carry out both technical and biological replicates. Direct quantification based on calibration curves with initial standards is tricky as it is difficult to obtain linear relation between standard concentrations and signal intensity due to partial recovery. We usually do at least three technical replicates,

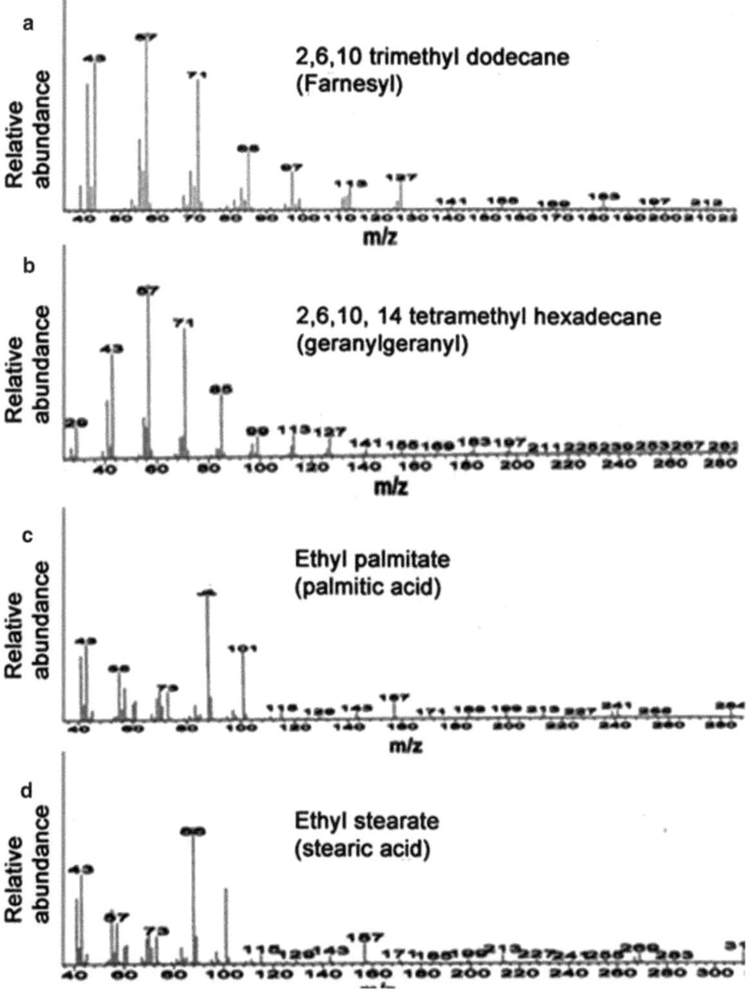

Fig. 3 MS data of the different lipid moieties. (**a**) MS data of 2,6′10 trimethyl dodecane, the reduced form of farnesyl. (**b**) 2,6,10,14 tetramethyl hexadecane, the reduced form of geranylgeranyl. (**c**) Ethyl palmitate, the reduced and ethyl modified form of palmitic acid. (**d**) Ethyl stearate, the reduced and ethyl modified form of stearic acid

analyzing the same lipid samples and three biological replicates, analyzing lipids samples derived from independent preparations of proteins. The identification of the lipid moieties is based on comparing the GC retention times of the sample with commercial initial standards that undergo similar treatment prior to their separation and comparison of the MS data at probability of $P \geq 95\%$. Complete MS data are shown in Fig. 3. As positive controls, we use authentic palmitic and stearic acids standards for the *S*-acylation analysis and *N*-acetyl *S*-farnesyl for the prenylation analysis. Negative controls include lipid preparations from the analyzed protein that was expressed in a different system, such as *E. coli*, point mutants in the lipid acceptor cysteine residues and protein preparations from cells that do not express the analyzed recombinant protein.

2 Materials

1. GC-MS Analysis program- Chemstation V.D.038 program (Agilent Technologies).
2. Bubbler.
3. Desiccator, glass desiccator only.
4. 100 % ethanol (HPLC/Spectrophotometric purity grade).
5. Formic acid.
6. 5 ml glass Vials for the Raney nickel cleavage (Wheaton ampoule, glass, amber, pre-scored, Sigma-Aldrich).
7. 30 ml glass vials for Raney nickel preparation (Kimble glass Inc., culture tube, Size/Cap 25 × 150 mm).
8. 7 × 2 mm magnetic stirring bar, PTFE.
9. Mass spectral Library-NIST98 (data version NIST 05, software version 2.0d).
10. Mercury.
11. *N*-Acetyl *S*-farnesyl-L-cysteine.
12. *N*-Acetyl-*S*-geranylgeranyl-L-cysteine.
13. Palmitic acid.
14. Pentane for GC.
15. Platinum (IV) oxide.
16. Raney nickel- ready to use.
17. Raney nickel—Aluminum–nickel alloy.
18. 15 mm TFE/Silicone Septa for glass V-vial (Wheaton, Sigma-Aldrich).
19. Sigmacoat SL2.
20. Sodium hydroxide.
21. Stearic acid.
22. 2 ml glass V-vial (Wheaton, Sigma-Aldrich).

3 Methods

The protocol consists of six major steps: (1) sample preparation, (2) sample wash, (3) hydrogenation, (4) Raney nickel cleavage, (5) lipid extraction and (6) GC-MS analysis. Steps 1–5 should be held in chemical hood.

3.1 Sample Preparation

A minimum of 25 μg of purified protein is required for the analysis of prenylation or 1 μg for *S*-acylation analysis. The protein concentrations need to be taken into account as the procedure involves lyophilization, which evaporates liquids but not salts. High salt concentrations may prevent solubilization of proteins after lyophilization. Typically, we maintain the proteins after purification in phosphate buffer containing 300 mM NaCl and 2 mM $MgCl_2$. It is recommended that total salt concentration would not be higher than 300 mM. The reaction is carried out in 2 ml glass V-vials. The glass V-vials need to be precoated with silicon to avoid adherence of lipids to the glass. The cap of the glass V-vials need to be covered with TFE/Silicon septa to avoid contact with plastic cap. From this step onward, it is essential to use only glass equipment because the pentane can solubilize the plastic and cause contaminations that would interfere with the identification in the GC-MS.

1. Coat the 2 ml glass V-vials with silicon to avoid adherence of lipids to the glass. Cover the cap with TFE/Silicon septa. Plastic often contaminates samples (because many components of plastic are soluble in organic solvents) compromising the identification in the GC-MS. Therefore, use glassware only.
2. Prepare 25–100 μg of purified protein for prenylation analysis or 1–25 μg for *S*-acylation analysis at a final concentration of 0.5–1.5 μg/μl in silicon-covered glass V-vials with TFE/Silicon septa inside the cap. The total salt concentration in the protein solubilization buffer should be around 300 mM. Higher salt concentrations may compromise solubilization of the protein.
3. Lyophilize protein samples until they are completely dry. A Speed-Vac could be used provided the glass V-vials fit into the rotor. Lyophilization usually takes up to 4 h but can be carried out overnight.
4. Resuspend the sample by adding 500 μl of **solution-I** (formic acid/absolute ethanol (1:4)) and mix with a vortex until the solution is clear (*see* **Note 1**). It is recommended to use vortex to ensure complete solubilization of the protein. Heating to 37 °C can be done to promote solubilization. The lyophilized protein powder is hygroscopic and therefore should be solubilized promptly.
5. As initial standards, 1 mg of each palmitic acid or stearic acid or *N*-acetyl *S*-farnesyl or *N*-acetyl *S*-geranylgeranyl is dissolved in 500 μl of **solution-I** (100 % ethanol/formic acid (4:1)). The initial standards should undergo the same procedure excluding the sample wash as this will wash away the standards.

3.2 Sample Wash

The next step is a sample wash with **solution-II** (**solution-I/** pentane (4:10)) to remove lipids that are not covalently bound to the protein. After phase separation the upper, pentane phase is discarded.

1. Wash the sample by adding 1 ml of **solution-II** to remove lipids that are not covalently attached to the protein. If no clear phases can be detected, add 200 μl of pentane and cool the samples to 4 °C. The low temperature should assist the phase separation (*see* **Note 2**).
2. Mix the samples with a vortex for 1 min and centrifuge at 1,500 × *g* for 5 min at 4 °C to separate the organic and aqueous phases. As the glass V-vials do not fit into our centrifuge rotors, we place them inside 50 ml plastic tubes and stabilize them with a paper towel.
3. Remove upper (organic) layer using a Pasteur pipette.
4. Repeat washes twice more with the same volumes of **solution-II**.

3.3 Hydrogenation

Hydrogenation is commonly used to reduce or saturate organic compounds and was shown to cause reduction of double bonds in diverse compounds and conditions [13]. The hydrogenation with platinum (IV) oxide in the presence of formic acid and ethanol, used in this protocol, causes reduction of the prenyl double bonds to insure single compound detection in the GC. A vacuum is applied to a glass desiccator using a pump. The hydrogen enters into the desiccator through the bubbler/trap system (a scheme of a typical hydrogenation system is presented in [9]). The bubbler should be filled with mercury so that the inner tube opening would be covered. Mineral oil could be used instead of mercury. For setup of the bubbler/trap, desiccator and vacuum pump: install the hydrogenation system such that from the trap, the hydrogen would flow through the mercury bubbler. The bubbler controls the hydrogen flow so that when the desiccator is filled, it allows the excess gas to exit from the system, preventing any possibility of overpressure. The bubbler is commercially available or can be made in-house by a glassware workshop. It is recommended to use a trap to prevent possible sucking of the mercury from the bubbler to the system due to backpressure that could occur should the vacuumed desiccator would be opened too fast. The trap/bubbler system is small and should fit into common chemical hoods in biology laboratories. Hydrogenation is commonly used in chemistry laboratories. Consult the relevant personnel at your institute on the required safety measures for installation of a hydrogen tank, pipeline and pressure controllers. We use hydrogen source with a pressure of 1 bar.

1. Add 5 mg of platinum (IV) oxide. Notice that bubbles will appear on addition of the platinum (IV) oxide to the protein solution.
2. Add a small magnet to the reactive glass V-vial.

3. Place the sample in the desiccator. The desiccator should be put on a magnetic stirrer. The sample must be stirred to ensure efficient hydrogenation.
4. Create a vacuum in the desiccator by operating the vacuum pump. In parallel, start the hydrogen flow in the bubbler/trap system. Make sure that hydrogen does not flow into the desiccator. Bubbles should form in the mercury inside the bubbler. Take care when creating vacuum to avoid evaporation of ethanol.
5. Hydrogen is highly volatile and may cause an explosion if not used properly. Consult your safety officer on the appropriate installation of the hydrogen source, pipeline and pressure controllers. Hydrogenation should be carried out in a chemical hood. Make sure that the hood evaporation system is turned on. Make sure that the bubbler is connected properly. Hydrogenation is commonly carried out by chemists and is not dangerous if safety measures are strictly followed.
6. Close the vacuum, disconnect the desiccator from the vacuum pump and flush the desiccator with hydrogen. Once the desiccator is filled with hydrogen gas bubbles will form in the bubbler. To make sure that no air is left and that the samples are kept under hydrogen, one can flush the desiccator twice more with hydrogen.
7. When the desiccator is filled with hydrogen gas, the bubbles will be forming in the bubbler. Close the desiccator and then close the hydrogen source.
8. Incubate the sample under hydrogen at 22–25 °C with continuous stirring for 2–4 h. Verify that the samples are well mixed during the process; otherwise the hydrogenation will be partial (*see* **Note 3**).
9. Stop the hydrogenation and pump the hydrogen with the vacuum pump. Make sure that the hydrogen source is closed before proceeding to **step 10**.
10. Disconnect the desiccator from the vacuum pump. Make sure that no bubbles are forming in the protein solution. If bubbles are still forming, the hydrogenation is not complete. In such case, repeat the hydrogenation for 1 h as described in **steps 3–9**. Samples can be stored at −20 °C for up to 24 h, without affecting the results much. However, shorter incubation periods are recommended.
11. For prenylation analysis continue to Subheading 3.4, Raney nickel cleavage. *For S-acylation analysis only, skip* Subheading 3.4 *and move directly to* Subheading 3.5, *lipid extraction.*

3.4 Raney Nickel Cleavage

This step is necessary only for prenylation analysis. The reduced prenyl groups are cleaved from the using the Raney nickel [6], which breaks down the thioether bond between the prenyl group and the protein [14]. Since ready-to-use Raney nickel reagent is often not commercially available, we will also describe how to prepare the Raney nickel reagent from Aluminum-nickel alloy. As with hydrogenation, preparation of the Raney nickel reagent and catalysis are commonly used in chemistry laboratories. Alternatively, Raney nickel ready to use can also be used.

The Raney nickel reagent is auto flammable when dry, extra caution should be taken when handling this reagent

1. To prepare the Raney nickel reagent mix 38 g of NaOH in 150 ml of deuterium-depleted water (DDW) in 500 ml glass beaker. Cool the solution to 4 °C on ice.
2. Add 15 g of the Aluminum-nickel alloy very slowly while stirring, make sure that the temperature does not reach 15 °C, and use extreme caution. This step should take approximately 3 h.
3. Transfer the beaker to 22 °C, bubbling will continue.
4. When the bubbling decreases, heat to 50 °C while stirring.
5. Incubate at 50 °C while stirring for 12 h. Make sure to add DDW to insure that the liquid level would not decrease.
6. Transfer the beaker to room temperature and let the alloy settle. Discard the solution.
7. Add 25 ml of DDW, shake gently and transfer the alloy to 30 ml glass vial.
8. Centrifuge for 1 min at 500 × *g*, discard the supernatant.
9. Prepare a solution of 5 g NaOH in 50 ml DDW and add 25 ml to the glass vial containing the alloy.
10. Shake well and incubate with the vial open at room temperature for 30 min.
11. Wash the alloy 40× times as follows: add 25 ml of DDW, shake briefly, centrifuge for 1 min at 500 × *g* and discard the supernatant.
12. Wash 5× times with 95 % Ethanol as follows: add 25 ml of 95 % Ethanol, centrifuge for 1 min at 500 × *g* and discard the supernatant.
13. Wash 10× times with 100 % ethanol as follows: add 25 ml of 100 % ethanol, centrifuge for 1 min at 500 × *g* and discard the supernatant. At the last wash make sure that the pH is around 7. If not, keep washing with 100 % ethanol until pH ~ 7.0.
14. After the last wash add 25 ml of 100 % ethanol. The Raney nickel reagent is now ready for use and can be stored up to 5 months at 4 °C in the dark.
15. To test the Raney nickel activity, weigh 10 mg on small piece of paper and let it dry for 5 min in the chemical hood. Since

active catalysis is auto flammable it will create a small fire. Make sure that you have a 500 ml chemical glass full of water next to you to stop the fire.

16. Transfer the solution from Subheading 3, **step 10** into 5 ml glass vials for the Raney nickel cleavage using a Pasteur pipette.
17. Using a spatula, add 50 mg of wet ready to use Raney nickel reagent from **step 14** (or Ready to use). Use extra caution not letting the Raney nickel reagent dries since it will ignite.
18. Add 100 μl of **solution I** (formic acid/absolute ethanol (1:4)).
19. Add small magnetic bar.
20. Incubate on dry ice for 10 min.
21. Seal the ampoule while it is on dry ice. It is strongly advisable that this step will be done by a professional glassblower (Chemistry departments often have such experts).
22. Incubate the mixture at 100 °C for 16 h while stirring.
23. Transfer the ampoule to dry ice and incubate for an hour.
24. Carefully open the ampoule by breaking the upper tip using small fretsaw. Use protective gloves and goggles since the ampoule still hold some pressure.
25. Add to the ampoule 400 μl of DDW.
26. Transfer all the solution into new 2 ml silicon precoated glass V-vials with septa inside the cap. Usually the platinum (IV) oxide and the Raney nickel reagents would be transferred as well. This should not interrupt the procedure.
27. At this stage all the reduce prenyl groups are released from the protein. The next steps will be separation of the lipid moieties (*see* **Note 4**).

3.5 Liquid/Liquid Extraction

When only analyzing *S*-acylation (in case when the procedure is continued directly from Subheading 3.3), add 400 μl of DDW to the hydrogenated protein sample to promote phase separation. The free lipids are then extracted into the pentane phase.

1. Extract the lipids by adding 0.5 ml of pentane.
2. Vortex for 1 min and centrifuge at 1,500 × *g* for 5 min at 4 °C. The low temperature is required for phase separation.
3. Take the upper organic layer, which contains the released lipids to new silicon-coated glass V-vial with septa inside the cap.
4. Repeat extraction twice more using the same volumes of pentane.
5. Collect all the lipid extraction into one sample of 1.5 ml.
6. Concentrate the 1.5 ml pentane lipid solution under a weak nitrogen flow to a final volume of 20 μl. Keep the glass V-vial

on ice and place the needle 5 cm above the sample. Make sure to adjust the N_2 flow before applying on the sample. Samples maybe stored up to 1 month at −20 °C without affecting the results.

7. Initial standards are concentrated under N_2 to a final volume of 100 μl.

3.6 GC-MS Analysis

Samples for analysis should be manually injected in 1 μl aliquots. In our work, we use a column with a 95 % dimethyl, 5 % diphenyl stationary phase and helium as the carrier gas. This column gives us reproducibly, well-separated narrow peaks of all the lipid moieties. Compounds are tentatively identified ($P \geq 95$ % match) on the basis of the NIST98 Mass Spectral Library (data version NIST 05, software version 2.0d) with the Chemstation V.D.038 program (Agilent Technologies). Further identification is based on comparisons of mass spectra and retention times with those of authentic standards analyzed under similar conditions. Ion 71 is a typical mass signature associated with prenyl lipids, and Ion 101 is a typical mass signature associated with acyl lipids. To discard contaminations, we can filter the GC data to ion 71 or 101. This procedure is commonly performed in GC-MS for clearer presentation of GC chromatograms and facilitating the analysis of single peaks. The GC retention time could differ from day to day. Therefore, it is essential to run farnesyl, geranylgeranyl, palmitic, and stearic acid standards alongside the samples.

1. On the GC-MS, set injection temperature to 250 °C (splitless mode), the interface to 280 °C and the ion source to 200 °C.
2. Inject 1 μl samples using manual injection.
3. Analyze samples using the following program: 5 min of isothermal heating at 100 °C, followed by a 5 °C/min oven temperature ramp up to 280 °C.
4. Equilibrated the system for 1 min at 100 °C before injection of the next sample.
5. Mass spectra are recorded at 4.59 scans per second with a 41–350 mass/charge (m/z) ratio scanning range and electron energy of 70 eV.
6. To discard contaminations, the GC data can be filtered to ions 71 or 101(*see* **Note 5**).
7. On the GC-MS each standard is first analyzed separately. Then, the two standards can be mixed and analyzed together. The standards are stable and can be kept at −20 °C for at least 1 year.

4 Notes

1. After lyophilization, sometimes the dried proteins do not dissolve in **solution-I**. This is due to high salt concentrations. Use dialysis to reduce salt concentrations. Avoid salt concentration higher than 300 mM.
2. During the wash step, sometimes there is no phase separation. This is due to Emulsion of **solution-I** with **solution-II**. Carry out the washes at 4 °C to promote phase separation and/or add 200 μl of pentane to promote phase separation.
3. Wide signals of the acyl groups in the GC chromatogram (usually, the MS data will contain the fatty acid molecular ion, *m/z* 256 for palmitic acid and *m/z* 284 for stearic acid) are due to inefficient hydrogenation. Make sure that the hydrogenation system works properly by testing the hydrogenation of palmitic acid and stearic acid standards.
4. Low signal of the prenyl groups can be due to inefficient Raney nickel cleavage. Use the initial standards *N*-acetyl *S*-farnesyl-L-cysteine and *N*-Acetyl-*S*-geranylgeranyl-L-cysteine to verify the cleavage efficiency.
5. Low signal in the GC chromatogram can be due to low quantities of the released lipid groups. Filter the ion chromatogram using ion 71 for prenylation and ion 101 for *S*-acylation and/or use larger amount of purified protein.

References

1. Wedegaertner PB, Wilson PT, Bourne HR (1995) Lipid modifications of trimeric G proteins. J Biol Chem 270:503–506
2. Adjobo-Hermans MJ, Goedhart J, Gadella TW Jr (2006) Plant G protein heterotrimers require dual lipidation motifs of Galpha and Ggamma and do not dissociate upon activation. J Cell Sci 119:5087–5097
3. Hemsley PA, Taylor L, Grierson CS (2008) Assaying protein palmitoylation in plants. Plant Methods 4:2
4. Sorek N, Gutman O, Bar E, Abu-Abied M, Feng X, Running MP, Lewinsohn E, Ori N, Sadot E, Henis YI, Yalovsky S (2010) Differential effects of prenylation and s-acylation on type I and II ROPS membrane interaction and function. Plant Physiol 155:706–720
5. Zeng Q, Wang X, Running MP (2007) Dual lipid modification of Arabidopsis Ggamma-subunits is required for efficient plasma membrane targeting. Plant Physiol 143:1119–1131
6. Farnsworth CC, Casey PJ, Howald WN, Glomest JA, Gelb MH (1990) Structural characterization of prenyl groups attached to proteins. Methods 1:231–240
7. Farnsworth CC, Wolda SL, Gelb MH, Glomset JA (1989) Human lamin B contains a farnesylated cysteine residue. J Biol Chem 264:20422–20429
8. Sorek N, Poraty L, Sternberg H, Bar E, Lewinsohn E, Yalovsky S (2007) Activation status-coupled transient S acylation determines membrane partitioning of a plant Rho-related GTPase. Mol Cell Biol 27:2144–2154
9. Sorek N, Yalovsky S (2010) Analysis of protein S-acylation by gas chromatography–coupled mass spectrometry using purified proteins. Nat Protoc 5:834–840
10. Thomas DW, van Kuijk FJ, Dratz EA, Stephens RJ (1991) Quantitative determination of hydroxy fatty acids as an indicator of in vivo lipid peroxidation: gas chromatography–mass

spectrometry methods. Anal Biochem 198: 104–111
11. Bernstein LS, Linder ME, Hepler JR (2004) Analysis of RGS protein palmitoylation. Methods Mol Biol 237:195–204
12. Drisdel RC, Alexander JK, Sayeed A, Green WN (2006) Assays of protein palmitoylation. Methods 40:127–134
13. Hudlický M (1996) Reductions in organic chemistry. American Chemical Society, Washington, DC
14. Uchida K, Stadtman ER (1992) Selective cleavage of thioether linkage in proteins modified with 4-hydroxynonenal. Proc Natl Acad Sci U S A 89:5611–5615

Chapter 14

In Vitro Myristoylation Assay of Arabidopsis Proteins

Xuehui Feng, Wan Shi, Xuejun Wang, and Mark P. Running

Abstract

Myristoylation is a lipid modification conserved among eukaryotes and involves the addition of a 14-carbon myristoyl moiety to a glycine at the N-terminus of cargo proteins. Since not every protein with an N-terminal glycine is myristoylated, experimental verification is necessary to determine which proteins are indeed myristoylated. Here we describe an in vitro myristoylation assay for the *Arabidopsis* heterotrimeric G protein alpha subunit, GPA1, as well as the *Arabidopsis* SALT OVERLY SENSITIVE3. This method can be easily adopted to other proteins of interest.

Key words Myristoylation, Myristoyl moiety, Arabidopsis, GPA1, SOS3

1 Introduction

Many eukaryotic proteins are subject to posttranslational modification. Myristoylation is a lipid modification that adds a 14-carbon myristoyl moiety to a glycine at the N-terminus of target proteins [1]. As with other lipid modifications, the primary functions of myristoylation is to facilitate membrane targeting and association as well as promote protein–protein interactions by virtue of the long hydrophobic chain. Myristoylation is conserved among Eukaryotes, including plants [2–4].

Among the earliest proteins shown to be myristoylated were mammalian heterotrimeric G protein alpha subunits (G_α); this lipid modification enhances its interaction with the G protein beta and gamma subunits ($G_{\beta\gamma}$) [5]. Heterotrimeric G proteins are eukaryotic signal transduction proteins that mediate a wide variety of cellular responses. In plants, heterotrimeric G proteins play roles in many developmental processes and environmental responses [6]. Plants have a single G_α subunit gene, *GPA1*, and, like in many other eukaryotes, it harbors a consensus sequence for myristoylation, including an N-terminal glycine. A peptide containing a portion of the N-terminal GPA sequence is myristoylated in vitro [3].

Mark P. Running (ed.), *G Protein-Coupled Receptor Signaling in Plants: Methods and Protocols*, Methods in Molecular Biology, vol. 1043, DOI 10.1007/978-1-62703-532-3_14, © Springer Science+Business Media, LLC 2013

In this chapter we describe an in vitro assay for myristoylation of full-length plant proteins. We demonstrate that full-length GPA1, along with an additional protein, SALT OVERLY SENSITIVE3 (SOS3) [7], is indeed myristoylated (Fig. 1). However, the assay can be adapted to any other protein of interest. The protocol is derived from [8].

2 Materials

1. Cloning vectors:
   - (a) pGEX-4T-1(N-terminal GST tag, GE Healthcare Bio-Sciences).
   - (b) pET21(C-terminal His tag, Novagen, *see* **Note 1**).
2. *E. coli* DH5α competent cells.
3. *E. coli* Rosetta/DE3 or BL21 protein expression strain.
4. Standard LB media.
5. Isopropyl-1-thio-β-D-galactopyranoside (IPTG).
6. Protein purification kits:
   - (a) Microspin GST Purification Module (Amersham Biosciences, 27-4570-03).
   - (b) His-Bind® Purification Kit (Novagen, 70239-3).
7. Isotope: [^{3}H] myristoyl-CoA (60 ci/mmol, 1 mci/ml).
8. 1× in vitro myristoylation buffer:
   - (a) 50 mM Tris–HCl (pH 7.8).
   - (b) 0.5 mM EGTA.
   - (c) 0.1 % Triton X-100.
9. 2× Protein sample buffer:
   - (a) 4 % SDS.
   - (b) 20 % glycerol.
   - (c) 10 % 2-mercaptoethanol.
   - (d) 0.004 % bromophenol blue.
   - (e) 0.125 M Tris–HCl (pH 6.8).
10. SDS-PAGE gel.
11. Gel fixing solution:
    - (a) 5 % acetic acid.
    - (b) 5 % isopropanol.
12. Autoradiographic image intensifier: Autofluor (National diagnostics, LS-315).

AtNMT1 has myristoyltransferase activity on full length proteins

GPA1 GPA1m SOS3 SOS3m

Fig. 1 In vitro myristoylation reaction of GPA1 and SOS3. GPA1 and SOS3 are myristoylated, while mutated proteins lacking the N-terminal glycine are not myristoylated

13. Ultra clear cellophane (Research Products International Corp, order No 1080).
14. X-ray films (KODAK BioMax XAR Film, East KODAK Co., Rochester, NY).

3 Methods

3.1 In Vitro Expression and Purification of AtNMT1 and GPA1 Proteins

1. Amplify *AtNMT1* and *GPA1* coding sequences by reverse transcription PCR using total RNA from *Arabidopsis* ecotype Columbia (Col-0).
2. Clone the *AtNMT1* fragment into pGEX-4T-1 vector.
3. Clone the *GPA1* fragment into pET-21b vector (*see* **Note 1**).
4. Transform pGEX-4T-1-*AtNMT1* and pET-21b-*GPA1* constructs into *E. coli* DH5α competent cells.
5. Screen colonies for correct plasmids.
6. Transform pGEX-4T-1-AtNMT1 and pET21b-GPA1 plasmids into *E. coli* Rosetta/DE3 strain (Novagen, Madison, WI) for protein expression.
7. Screen for positive colonies and inoculate in 5 ml LB broth.
8. After shaking at 37 °C overnight, inoculate a bigger volume culture (100–250 ml) with the 5 ml culture and incubate until an A_{600} of 0.6–0.8 (3–5 h).
9. Induce protein expression by adding 100 μM isopropyl-1-thio-β-D-galactopyranoside (IPTG) for the pET21b construct or 500 μM IPTG for the pGET-4T-1 construct, and continue the incubation for an additional 1–2 h at 30 °C.
10. Purify recombinant *AtNMT1* protein using Microspin GST Purification Module (Amersham Biosciences, Buckinghamshire, England) following the manufacturer's instructions.
11. Purify recombinant *GPA1* protein using His·Bind® Purification Kit (Novagen, Madison, WI) following the manufacturer's instructions.
12. Store purified *AtNMT1* and *GPA1* protein aliquots at −80 °C.

3.2 GPA1 In Vitro Myristoylation Assay

The in vitro myristoylation assay is performed essentially as described [8], and includes the following steps:

1. Add purified *GPA1* and *AtNMT1* proteins (100–500 ng each) to the myristoylation reaction buffer, containing 50 mM Tris–HCl (pH 7.8), 0.5 mM EGTA, and 0.1 % Triton X-100.
2. Add 0.4 μM (0.6 μl) [^{3}H] myristoyl-CoA (60ci/mmol, 1 mci/ml) to initiate the reaction. The final volume is 25 μl.
3. Incubate the myristoylation reaction at 30 °C for 10–30 min.

4. Terminate the reaction by adding 8 μl 4× protein loading dye and heat the tube at 95 °C for 5 min.
5. Spin for 1 min and load 15 μl of the solution to a 10 % SDS-PAGE gel to separate proteins and reagents.
6. After electrophoresis, fix the gel in gel fixing solution, containing 5 % acetic acid and 5 % isopropanol, for 30 min with shaking.
7. Rinse the gel thoroughly for 15 min with tap water (*see* **Note 2**).
8. Amplify tritium signal in the gel using Autoradiographic Image Intensifier solution (Autofluor: National diagnostics, LS-315) with 0.5 % glycerol, and shake for 30 min.
9. Wet a piece of ultra clear cellophane (Research Products International Corp, order No 1080) in Autoradiographic Image Intensifier solution.
10. Cover the gel with cellophane without washing.
11. Dry the gel on gel dryer at 80 °C for 1–2 h.
12. Expose the gel to X-ray films (KODAK BioMax XAR Film, East KODAK Co., Rochester, NY) at −80 °C for 4–5 days.

4 Notes

1. Use C-terminal, but not N-terminal tag vectors to express myristoylation target proteins, to avoid blocking the N-terminal glycine.
2. Follow appropriate radiation regulations.

References

1. Farazi TA, Waksman G, Gordon JI (2001) The biology and enzymology of protein N-myristoylation. J Biol Chem 276: 39501–39504
2. Qi Q, Rajala RV, Anderson W, Jiang C, Rozwadowski K, Selvaraj G, Sharma R, Datla R (2000) Molecular cloning, genomic organization, and biochemical characterization of myristoyl-CoA:protein N-myristoyltransferase from *Arabidopsis thaliana*. J Biol Chem 275: 9673–9683
3. Boisson B, Giglione C, Meinnel T (2003) Unexpected protein families including cell defense components feature in the N-myristoylome of a higher eukaryote. J Biol Chem 278:43418–43429
4. Boisson B, Meinnel T (2003) A continuous assay of myristoyl-CoA:protein N-myristoyltransferase for proteomic analysis. Anal Biochem 322: 116–123
5. Linder ME, Pang IH, Duronio RJ, Gordon JI, Sternweis PC, Gilman AG (1991) Lipid modifications of G protein subunits. Myristoylation of Go alpha increases its affinity for beta gamma. J Biol Chem 266:4654–4659
6. Temple BR, Jones AM (2007) The plant heterotrimeric G-protein complex. Annu Rev Plant Biol 58:249–266
7. Liu J, Zhu JK (1998) A calcium sensor homolog required for plant salt tolerance. Science 280:1943–1945
8. Selvakumar P, Lakshmikuttyamma A, Sharma RK (2009) Biochemical characterization of bovine brain myristoyl-CoA:protein N-myristoyltransferase type 2. J Biomed Biotechnol 2009:907614

Chapter 15

Assaying Protein S-Acylation in Plants

Piers A. Hemsley

Abstract

S-acylation is increasingly being recognized as an important posttranslational modification of proteins controlling activity, subcellular localization, microdomain residence, and stability. Heterotrimeric G-proteins and GPCRs are particularly well studied S-acylated proteins, and fast, cheap, reliable methods are required for the analysis of S-acylation states of these proteins. Various approaches have been developed to study S-acylation, but they are time consuming, expensive, frequently require radiolabels and generally only suitable for cell culture, making them impractical for work in plant systems. Here a rapid and inexpensive method is described for the analysis of the S-acylation state of AGG2 that can be performed on any cell or tissue sample using standard laboratory equipment and methods. This method is also applicable to any protein that can be detected by western blotting.

Key words AGG2, S-acylation, Palmitoylation, S-acylated, Palmitoylated, Heterotrimeric G-protein

1 Introduction

Heterotrimeric G-proteins are ubiquitous signaling molecules in eukaryotes conveying signals within the cell after stimulation of G-protein coupled receptors (GPCRs). Posttranslational modification of GPCRs and heterotrimeric G-protein subunits by lipid groups such as palmitate (S-acylation or palmitoylation) is well documented [1–3]. These modifications can increase affinity of a molecule for the membrane but can also affect the sensitivity of a GPCR or the subcellular localization of heterotrimer components [2, 3]. Arabidopsis contains one Gα (GPA1), one Gβ (AGB1) and three Gγ (AGG1, AGG2 and AGG3) heterotrimeric G-protein subunits. GPA1 [2] and AGG2 [2, 4, 5] are thought to be S-acylated. Mutation of putative S-acylation sites in GPA1 or AGG2 leads to their mislocalization away from the plasma membrane, but mutation of a similar putative site in AGG1 appeared to have little effect [2, 3].

Traditionally, analysis of S-acylation has involved radiolabelling of cell cultures using tritiated fatty acids such as ^{3}H-palmitic acid

Mark P. Running (ed.), *G Protein-Coupled Receptor Signaling in Plants: Methods and Protocols*, Methods in Molecular Biology, vol. 1043, DOI 10.1007/978-1-62703-532-3_15, © Springer Science+Business Media, LLC 2013

followed by immunoprecipitation, SDS-PAGE and autoradiography. This method, although useful for mammalian, yeast and plant cell culture research, is not suitable to the study of proteins being analyzed in whole plant tissues, situations where proteins are expressed at low level or not expressed in cell culture and is very expensive and time consuming. Recently a faster method based on the substitution of biotin for S-acyl groups [6] removed the need for these costly and, for plant research, impractical methods. Subsequent work has optimized the assay to work with any epitope [5, 7] allowing it to be used on any protein that can be detected by western blot. Here a method is presented to analyze the S-acylation state of AGG2. This method has also been successfully used to analyze and demonstrate the S-acylation of many other plant proteins, including the heterotrimeric G-protein subunits GPA1 [8].

2 Materials

2.1 Plant Material

1. Plant tissue expressing AGG2, preferably to levels detectable by western blot of 50 μg total protein or membrane preparation.

2.2 Solutions

1. 1 M Hydroxylamine in water pH 7.4 (pH with NaOH, prepare fresh).
2. 4 mM Biotin-HPDP (Thermo Fisher) in dimethylformamide (DMF) (prepare fresh).
3. Phosphate buffered saline (PBS) pH 7.4.
4. Lysis buffer: 1× PBS pH 7.4, Protease inhibitors, 5 mM EDTA, 1 % Triton X-100, 25 mM N-ethylmaleimide (prepare fresh, *see* **Notes 1** and **2**).
5. Resuspension buffer: 1× PBS pH 7.4, 8 M Urea, 2 % SDS. (store at −20 °C in 1 ml aliquots).
6. Wash buffer: 1× PBS pH 7.4, NaCl to 500 mM, 0.1 % SDS.
7. Methanol.
8. Chloroform.

2.3 Equipment and Consumables

1. 1.5 ml microfuge tubes.
2. 15 ml falcon tubes.
3. Microfuge.
4. Centrifuge capable of spinning 15 ml falcon tubes >5,000 ×*g*.
5. Nutating table.
6. Sonicating water bath (optional).
7. Equipment and reagents for running and blotting SDS-PAGE gels.

8. Miracloth.
9. Neutravidin agarose beads (Thermo Scientific).
10. Equipment and reagents for performing western blots.

3 Methods

The quantity of tissue used depends upon the level of expression of AGG2 in the sample and the expected level of S-acylation. To perform the required steps on a small scale for analysis by western blot it is best not to use more than 1 mg of starting protein. 500 mg of plant tissue is sufficient for AGG2 expressed at high level (e.g., 35S driven constructs in Arabidopsis), but if membrane purification to enrich for AGG2 is required then >2 g of plant tissue should be used to obtain sufficient starting protein. The protocol is essentially split into extraction, blocking, biotin substitution of S-acyl groups, purification on neutravidin beads and western blotting, and the protocol may be paused after each of these steps.

3.1 Assay Using Whole Plant Material

1. Grind snap-frozen plant tissue in liquid nitrogen to a fine powder and resuspend in 500 μl lysis buffer (*see* **Notes 1** and **2**) on ice.
2. Mix gently at 4 °C for 10 min.
3. Centrifuge at 4 °C, 500 × *g* for 10 min to remove insoluble material.
4. Determine protein concentration using a protein concentration assay compatible with 1 % Triton X −100.
5. Samples may be flash frozen and stored at −80 °C until needed.

3.2 Assay Using Purified Membranes

1. Grind >2 g snap-frozen plant tissue in liquid nitrogen to a fine powder and resuspend in 10 ml lysis buffer *lacking* Triton X-100 (*see* **Note 2**) on ice.
2. Filter through two layers of Miracloth.
3. Purify the membranes by ultracentrifugation at >65,000 × *g* for 1 h at 4 °C.
4. Gently resuspend the membranes in 200 μl lysis buffer with 1 % Triton X-100 (*see* **Note 1**).
5. Spin at 10,000 × *g* for 10 min at 4 °C to remove insoluble debris.
6. Determine protein concentration using an assay compatible with 1 % Triton X-100 (e.g.: Bio-Rad DC protein assay).
7. Samples may be snap-frozen and stored at −80 until needed.

3.3 Blocking

1. Take 1 mg of protein and dilute in lysis buffer to a final volume of 1 ml in a 15 ml falcon tube.
2. Incubate with gentle mixing at 4 °C for 12 h (*see* **Note 3**).

3. Precipitate proteins using chloroform/methanol [9]. Add 1 vol chloroform to sample, mix, add 3 vols methanol, mix, add 4 vols water, mix.
4. Centrifuge at >5,000 × *g* for 30 min at room temperature.
5. Remove supernatant taking care not to disturb the interphase.
6. Add 4 vols. of methanol and mix.
7. Incubate at −20 °C for 1 h to overnight (convenient stopping point).

3.4 Biotin Substitution of S-Acyl Groups

1. Centrifuge at >5,000 × *g* for 30 min at 4 °C.
2. Remove supernatant and allow to briefly air dry (Do not over dry).
3. Resuspend the pellet in 200 μl of resuspension buffer by sonication in a sonicating water bath for 10 min (optional) and gentle agitation on a roller table at room temperature until solubilized (*see* **Note 4**).
4. Divide the solution into two equal aliquots and combine one aliquot (experimental sample: Hyd+) with 200 μl PBS, 600 μl of 1 M fresh hydroxylamine solution, 1 mM EDTA, protease inhibitors, and 100 μl fresh 4 mM biotin-HPDP dissolved in DMF. Treat the remaining (negative control: Hyd−) aliquot identically but replace hydroxylamine solution with water (*see* **Note 5**). Gently mix for 1 h at RT in 15 ml falcon tubes on a nutating table.
5. Precipitate proteins at room temperature using methanol/chloroform as described previously (convenient stopping point).

3.5 Purification on Neutravidin Beads

1. Resuspend each sample in 100 μl of resuspension buffer as described previously and add 900 μl PBS containing 0.2 % Triton X-100. Mix gently.
2. Centrifuge at 5,000 × *g* for 10 min to pellet insoluble debris.
3. Remove 100 μl from each sample to act as a loading control (Hyd+ LC and Hyd− LC) and chloroform/methanol precipitate as described previously and resuspend in 25 μl 2× SDS-PAGE sample buffer containing 40 % v/v glycerol.
4. Combine the remaining 900 μl samples (Hyd+ EX and Hyd− EX) with 15 μl of high capacity neutravidin-agarose beads in a 1.5 ml microfuge tube for 1 h at room temperature on a roller table. Avoid transferring any pellet from **step 2**.
5. Collect the neutravidin beads by centrifugation at 1,000 × *g*, discard the supernatant and wash with 1 ml wash buffer. Repeat twice.
6. Collect the beads and wash with 1 ml PBS.

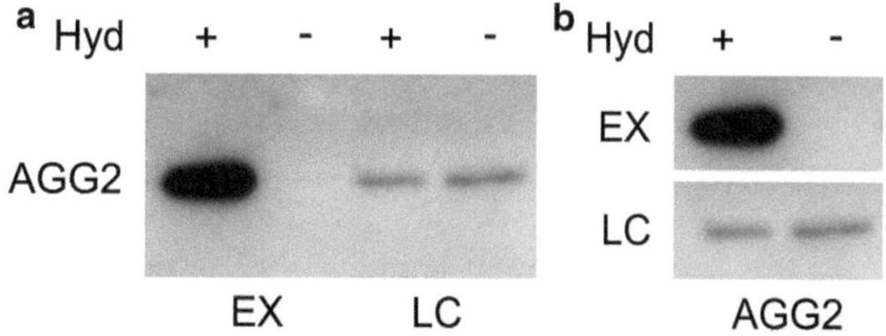

Fig. 1 AGG2 is S-acylated. (**a**) A signal in the experimental (EX) hydroxylamine (Hyd+) treated sample lane but not in the EX Hyd− sample lane indicates S-acylation of AGG2. Equal loading control band intensities confirms that the observed signal in the EX lanes is not due to bias in the experimental procedure. LC samples are 10 % of the total protein and EX samples are representative of the remaining 90 % after purification on neutravidin beads. (**b**) An alternative way of presenting the data in (**a**) allowing for more samples to be presented side by side in a more intuitive manner

7. Collect the beads and elute captured proteins in 25 μl of 2× SDS sample elution buffer containing 40 % glycerol v/v and 1 % 2-mercaptoethanol v/v at 95 °C for 5 min (convenient stopping point before heating step, store samples at −20 °C until required).

3.6 Analyze Samples by SDS/PAGE and Western Blotting

1. Run samples in the order shown in Fig. 1 on a denaturing 15 % SDS-PAGE gel.
2. Use standard/local methods to blot and detect AGG2.

3.7 Interpretation of Western Blot

1. For AGG2 to be considered S-acylated signal should be detected in the EX Hyd+ lane but not in the EX Hyd− lane (*see* Fig. 1). Equal signal should be detected in the LC lanes (*see* **Note 6**).

4 Notes

1. Lysis buffer containing up to 1 % SDS instead of Triton X-100 can also be used if poor blocking is observed or protein yield is low. Use of SDS can alter S-acylation profiles when the S-acylation state is dependent on native protein conformation, protein–protein interactions, etc. 0.5–1 % Saponin in addition to Triton X-100 can also be useful if cholesterol enriched regions of the membrane need to be disrupted.
2. N-ethylmaleimide should be prepared fresh before each use as a 100 mM stock in PBS supplemented with 5 mM EDTA.
3. The blocking step can be carried out overnight at 4 °C and provides a useful stopping point.

4. Gentle mixing at 37 °C for 10 min can improve solubilization of proteins.
5. Sometimes it is difficult to maintain the pH of hydroxylamine solutions during the assay leading to protein cleavage and non-specific labelling. In these circumstances make up hydroxylamine in 50 mM Tris–HCl and substitute the water in the negative control for 50 mM Tris–HCl pH 7.4.
6. If no signal is observed in the EX lanes this indicates that AGG2 is not S-acylated. If signal is observed in both EX lanes, then it is most likely that blocking was incomplete. If no signal is observed in either or both of the LC lanes, then either AGG2 is not expressed highly enough or sample was lost during processing. In either of these latter cases no useful data can be extracted and the experiment must be repeated.

References

1. Ovchinnikov YA, Abdulaev NG, Bogachuk AS (1988) Two adjacent cysteine residues in the C-terminal cytoplasmic fragment of bovine rhodopsin are palmitylated. FEBS Lett 230: 1–5
2. Adjobo-Hermans MJ, Goedhart J, Gadella TW Jr (2006) Plant G protein heterotrimers require dual lipidation motifs of Galpha and Ggamma and do not dissociate upon activation. J Cell Sci 119:5087–5097
3. Zeng Q, Wang X, Running MP (2007) Dual lipid modification of Arabidopsis Ggamma-subunits is required for efficient plasma membrane targeting. Plant Physiol 143:1119–1131
4. Running MP, Lavy M, Sternberg H, Galichet A, Gruissem W, Hake S, Ori N, Yalovsky S (2004) Enlarged meristems and delayed growth in plp mutants result from lack of CaaX prenyltransferases. Proc Natl Acad Sci USA 101: 7815–7820
5. Hemsley PA, Taylor L, Grierson CS (2008) Assaying protein palmitoylation in plants. Plant Methods 4:2
6. Drisdel RC, Green WN (2004) Labeling and quantifying sites of protein palmitoylation. Biotechniques 36:276–285
7. Hou H, Subramanian K, LaGrassa TJ, Markgraf D, Dietrich LE, Urban J, Decker N, Ungermann C (2005) The DHHC protein Pfa3 affects vacuole-associated palmitoylation of the fusion factor Vac8. Proc Natl Acad Sci USA 102: 17366–17371
8. Hemsley PA, Weimar T, Lilley KS, Dupree P, Grierson CS (2013) A proteomic approach identifies many novel palmitoylated proteins in Arabidopsis. New Phytol 197:805–814
9. Wessel D, Flugge UI (1984) A method for the quantitative recovery of protein in dilute solution in the presence of detergents and lipids. Anal Biochem 138:141–143

Chapter 16

In Vitro Prenylation Assay of Arabidopsis Proteins

Wan Shi, Qin Zeng, and Mark P. Running

Abstract

Protein prenylation, like other lipid posttranslational modifications of eukaryotic proteins, plays important roles in protein–membrane association and protein–protein interactions. In *Arabidopsis*, hundreds of proteins involved in a great variety of biological processes are potential prenylation substrates that need to be verified, including heterotrimeric G proteins and most Rop and Rab small GTPases. Also, genetic evidence suggests substrate cross-specificity among different prenyltransferases and/or the existence of unidentified prenylation players. In this chapter we describe a direct and flexible in vitro enzymatic assay designed for testing prenylation activity and substrate specificity in vitro. This protocol takes *Arabidopsis* Rab-GGT as example and starts with preparation of purified protein components of the reaction, followed by reconstitution of the prenylation reaction in vitro, and autoradiographic detection for qualitative and semiquantitative analysis.

Key words Protein expression and purification, Protein prenylation, In vitro enzymatic assay, Rab GTPase, Rab geranylgeranyltransferase, Rab escort protein, Autoradiography

1 Introduction

Protein prenylation is one of the posttranslational lipid modifications of eukaryotic proteins, and involves covalently adding a 15-carbon farnesyl or 20-carbon geranylgeranyl chain to the C-terminal cysteine residues of target proteins. Like other lipid modifications, prenylation plays important roles in facilitating protein–membrane association and protein–protein interactions [1].

In *Arabidopsis*, three types of heterodimeric enzymes, called protein prenyltransferases, catalyze three types of prenylation reactions: protein farnesyltransferase (PFT) adds one farnesyl moiety, protein geranylgeranyltransferase-I (PGGT-I) adds one geranylgeranyl moiety, and Rab geranylgeranyltransferase (Rab-GGT) adds one or two geranylgeranyl moieties to the target proteins [2–5]. Based on the C-terminal sequence that prenyltransferases recognize, at least 700 *Arabidopsis* proteins meet the minimal criteria of being potentially prenylated. These potential prenylation substrates are involved in a great variety of biological processes [6, 7].

Mark P. Running (ed.), *G Protein-Coupled Receptor Signaling in Plants: Methods and Protocols*, Methods in Molecular Biology, vol. 1043, DOI 10.1007/978-1-62703-532-3_16, © Springer Science+Business Media, LLC 2013

To address the role of prenylation in the biological function of potential target proteins, an enzymatic assay is needed to first test if they can be prenylated by any of the known prenyltransferases.

Several of the prenyltransferases have been characterized genetically. The α subunit that is shared between PFT and PGGT, termed *PLURIPETALA* (*PLP*), shows severe meristem and flower patterning defects and ABA sensitivity, but is nevertheless viable and fertile [8]. The β subunit of PFT, *ENHANCED RESPONSE TO ABSCISIC ACID1* (*ERA1*), also shows severe ABA sensitivity but only mild developmental defects, while the β subunit of PGGT, termed *GGB*, has no phenotype under normal growing conditions [9–11]. Of the two Rab-GGT α and two Rab-GGT β subunits in *Arabidopsis*, only one β subunit, *RGTB1*, has been characterized genetically, and has defects in apical dominance and gravitropism [12]. The viability and fertility of the *Arabidopsis* PFT/PGGT-I mutant is strikingly opposite to its animal counterpart, suggesting that *Arabidopsis* prenyltransferases have more cross specificity on substrates, and/or additional prenylation players exist but remain unidentified [7, 13]. An enzymatic assay is also very useful for testing the substrate specificity of the known prenyltransferases, as well as identifying new prenylation players.

In vitro prenylation assays provide a direct, flexible, and relatively easy way to detect prenyltransferase activity and determine substrate specificity. The assay involves reconstituting the prenylation reaction in a test tube, by incubating purified prenyltransferase, substrate protein, labeled prenyl moiety, and additional cofactors under optimized reaction conditions. If the prenyltransferase can catalyze the attachment of the labeled prenyl moiety to the substrate protein, the substrate protein will be distinguished from unprenylated proteins by carrying the label for detection.

This chapter takes one of the reactions catalyzed by *Arabidopsis* Rab-GGT as example to describe the materials, techniques, and procedures of in vitro prenylation assays. The Rab-GGT α subunit RGTA1 and β subunit RGTB1 are co-expressed in yeast, and co-purified in heterodimeric form by pull-down to ensure the enzymatic activity. Multiple *Arabidopsis* Rab small GTPases are reported to be prenylated by Rab-GGT [12, 14], among which RabA4b is chosen, expressed in and purified from *E. coli*. Rab escort protein (REP) is thought to be a required cofactor for Rab-GGT activity at least in rat [14, 15]; thus, *Arabidopsis* REP (AtREP) is also expressed in and purified from *E. coli* for the assay. The geranylgeranyl moiety used in the assay is radiolabeled with ^{3}H and can be detected by various methods such as autoradiography and scintillation counting. The reaction buffer also provides ions shown to be required for the Rab-GGT activity, as well as detergent that helps maintain protein–protein interactions.

2 Materials

2.1 Reagents and Solutions

1. LB broth:
   10 g/L tryptone, 5 g/L yeast extract, 10 g/L NaCl, pH 7.0, autoclaved.
2. 2× YTA broth, autoclaved:
   16 g/L tryptone, 10 g/L yeast extract, 5 g/L NaCl, pH 7.0, autoclaved.
3. 100 mg/mL Carbenicillin water solution, filter-sterilized.
4. 100 mM IPTG water solution, filter-sterilized.
5. 10 mM Phosphate buffered saline (PBS), pH 7.4:
   10 mM Na_2HPO_4, 140 mM NaCl, 2.7 mM KCl,1.8 mM KH_2PO_4, pH 7.4, autoclaved.
6. 20 mg/mL lysozyme (Sigma-Aldrich, L7651) water solution.
7. 1 M DTT water solution, filter-sterilized.
8. 10 μg/mL DNase I solution:
   10 μg/mL DNase I (Sigma-Aldrich, D4527), 10 mM Tris–HCl pH 7.5, 50 mM NaCl, 10 mM $MgCl_2$, 1 mM DTT, 50 % (v/v) glycerol, filter-sterilized.
9. GST SpinTrap columns (GE Healthcare, 28-9523-59).
10. Glutathione elution buffer:
    10 mM L-glutathione reduced (Sigma-Aldrich, G4251), 50 mM Tris–HCl pH 8.0
11. Novagen His-Bind Purification Kit (EMD Millipore, 70239–3)
12. Synthetic Dextrose–histidine (SD-His) dropout media:
    1.7 g/L Difco yeast nitrogen base without amino acids and ammonium sulfate (BD Diagnostics, 233520), 0.77 g/L CSM-His single dropout amino acids mixture powder (MP Biomedical, 114510312), 5 g/L $(NH_4)_2SO_4$, 20 g/L glucose, autoclaved (15 min).
13. Synthetic Galactose–histidine (SG-His) dropout media:
    The recipe is same as that of SD-His listed above, except using galactose instead of glucose as the carbon source.
14. CelLytic Y Plus Kit (Sigma-Aldrich, CYP1-1KT).
15. EZview Red FLAG M2 affinity gel (Sigma-Aldrich, F2426).
16. 50 mM Tris buffered saline (TBS), pH 7.4:
    50 mM Tris–HCl, 150 mM NaCl, pH 7.4, autoclaved.
17. 3× FLAG peptide stock solution:
    5 μg/μL 3× FLAG peptide, 100 mM Tris–HCl pH 7.5, 200 mM NaCl, filter-sterilized.

18. 5× Rab-GGT reaction buffer [16]:
    250 mM HEPES-KOH pH 7.8, 25 mM $MgCl_2$, 250 μM $ZnCl_2$, 1.5 % NP-40, 25 mM DTT, filter-sterilized.
19. Diafiltration buffer:
    50 mM HEPES-KOH pH 7.5, 1 mM DTT, filter-sterilized.
20. 30 Ci/mmol, 1 mCi/mL, all trans-Geranyl geranyl pyrophosphate [1-^{3}H] triammonium salt ([^{3}H]-GGPP, American Radiolabeled Chemicals, ART0348).
21. 6× Protein sample loading buffer:
    378 mM Tris–HCl pH 6.8, 30 % (v/v) glycerol, 12 % (w/v) SDS, 0.015 % (w/v) bromophenol blue, 6 % (v/v) β-mercaptoethanol (Sigma-Adrich, M3148).
22. 10× SDS-PAGE running buffer:
    250 mM Tris, 1.92 M glycine, 1 % (w/v) SDS
23. 10 % SDS-PAGE mini gel, 1 mm, e.g., Mini-PROTEAN TGX precast gel (Bio-Rad, 456–1034).
24. Gel fixation buffer:
    5 % (v/v) isopropanol, 5 % (v/v) acetic acid.
25. 50 % (v/v) glycerol, autoclaved.
26. Autofluor autoradiography intensifier (National Diagnostics, LS-315).
    Store 1, 2, 12, 13, 16, 22, 24, 25, 26 at room temperature; 5, 9,11, 23 at 4 °C; 3, 4, 6, 7, 8, 10, 14, 15, 17, 18, 19, 20, 21 at −20 °C.

2.2 Constructs and Cell Strains

1. The coding sequence of *RGTB1* is cloned into the multiple cloning site #1 of the pESC-HIS vector (Agilent Technologies), fused with a C-terminal FLAG epitope tag. The coding sequence of *RGTA1* is subsequently cloned into the multiple cloning site #2 of the intermediate pESC-HIS RGTB1-FLAG vector, fused with a C-terminal c-Myc epitope tag. The resulting pESC-HIS RGTB1-FLAG RGTA1-cMyc construct is transformed into yeast expression strain YPH499 (Agilent Technologies).
2. The coding sequence of *AtREP* is cloned into pET21b vector (Novagen), fused with a C-terminal His-tag. The resulting pET21-REP construct is transformed into *E. coli* expression strain BL21.
3. The coding sequence of *RabA4b* is cloned into the pGEX-4T-1 vector (GE Healthcare), fused with an N-terminal GST tag. The resulting pGEX-RabA4b construct is transformed into *E. coli* expression strain BL21.

2.3 Equipment

1. Centrifuge and rotors for 50 mL centrifugal tubes and 250 mL centrifugal bottles.
2. 250 mL, 500 mL flasks, 50 mL centrifugal tubes, 250 mL centrifugal tubes.
3. Sonicator with appropriate tip.
4. Water bath: 30, 37, 98 °C.
5. Roller drum and platform shaker.
6. Glass beads, acid washed, 425–600 μm (Sigma-Aldrich, G8772).
7. Vortex-based bead beater.
8. Amicon Ultra-0.5 mL centrifugal filters from protein purification and concentration, Ultracel-10 Membrane (Millipore, UFC501024).
9. UV/Vis spectrophotometer and disposable cuvettes.
10. Bio-Rad Mini-PROTEAN electrophoresis cell (Bio-Rad) and compatible power source.
11. Gel dryer and corresponding vacuum pump, with a cold-trap set up in between according to the requirements and recommendations by radiation safety office.
12. Kodak BioMax XAR film, 13 cm × 18 cm (Kodak, 1651496) and autoradiography cassette.
13. Filter paper and plastic wrap.
14. X-ray film developer.

3 Methods

Manipulations should be performed on ice or at 2–8 °C, unless otherwise specified.

3.1 Preparation of GST-Tagged RabA4b Protein

This subheading describes the procedure of purifying up to 1 mg of GST-tagged *Arabidopsis* Rab protein from a 40 mL *E. coli* culture using GST SpinTrap columns, which follows the manufacturer's protocol in general, with a number of adaptations [17].

1. Inoculate a single colony or a small portion of glycerol stock of BL21 pGEX-RabA4b strain in 5 mL of LB broth with 50 μg/mL carbenicillin. Shake overnight at 37 °C.
2. Inoculate 800 μL of the overnight culture into 40 mL fresh 2× YTA broth with 50 μg/mL carbenicillin (1:50 ratio). Shake at 37 °C for about 1.5 h, until the OD_{600} reaches 0.6–0.8.
3. Add 160 μL of 100 mM IPTG to a final concentration of 0.4 mM. Shake at room temperature to induce the protein expression for 5–6 h (*see* **Note 1**).

4. Harvest the cells by centrifugation at 5,000 × *g* for 5 min. Decant the supernatant and resuspend the pellet in 2 mL of 1× PBS.
5. Add 10 μL of 20 mg/mL lysozyme solution, 2 μL of 10 mg/mL DNase I and 10 μL of 1 M DTT to the suspension. Mix gently and incubate at room temperature for 5 min.
6. Disrupt the cells by sonication. The exact parameters are not specified here since they may vary a lot among different models of sonicators (*see* **Note 2**).
7. Clarify the lysate by centrifugation at 12,000 × *g* for 10 min. Transfer the supernatant to a fresh tube.
8. Resuspend the resin in a GST SpinTrap column by vortexing gently. Loosen the top cap of the column by 1/4 turn. Remove and save the bottom cap.
9. Place the column in a 2 mL collection tube and spin at 735 × *g* for 1 min. Discard the storage buffer in the collection tube.
10. Add 600 μL of the supernatant in **step 7** to the column, tighten the top cap and replace the bottom cap.
11. Mix the supernatant with resin by shaking gently (a roller drum will be very handy) at room temperature for 5 min.
12. Loosen the top cap and remove (and save) the bottom cap before place the column back to collection tube. Spin at 735 × *g* for 1 min. Discard the flow through or collect for analysis.
13. Repeat **steps 10–12** until all supernatant has flowed through the column.
14. Add 600 μL of 1× PBS to the column and replace the caps. Vortex briefly and spin as above to wash the resin. Discard the flow through. Repeat the wash once for a total of two washes.
15. Add 150 μL of glutathione elution buffer to the column and replace the caps. Incubate at room temperature for 7 min with gentle shaking.
16. Loosen the top cap and remove the bottom cap before placing the column in a clean 1.5 mL microcentrifuge tube. Spin at 735 × *g* to collect the elute. Repeat the elution procedure once and combine the two elutes to a total volume of 300 μL. The final elute can be temporarily stored at 4 °C for dialysis and quantification (*see* **Note 3**).

3.2 Preparation of His-Tagged AtREP Protein

This subheading describes the procedure of purifying up to 1 mg His-tagged AtREP protein from a 40 mL *E. coli* culture using the His-Bind purification kit, which follows the manufacturer's protocol in general, with a number of adaptations [18].

1. Follow the same procedure described in **steps 1–7** under Subheading 3.1 to grow BL21 pET21-AtREP cells, induce the protein expression, and make the cell lysate by sonication, except that do not add DTT in **step 5** (*see* **Note 4**).

2. Transfer 200 μL of 50 % slurry of His-Bind resin (part of the His-Bind purification kit) to a 1.5 mL microcentrifuge tube. Spin at 1,000 × *g* for 1 min. Remove the supernatant with a pipette.
3. Charge and equilibrate the resin by a series washes in the following sequence:
   (a) 200 μL sterilized water, twice.
   (b) 200 μL of 1× charge buffer (part of the His-Bind kit), three times.
   (c) 200 μL of 1× binding buffer (part of the His-Bind kit), twice.

   For each wash, add the buffer, mix by vortex gently, spin at 1,000 × *g* for 1 min, and remove the supernatant with a pipette.
4. Add 1 mL of the cell lysate from **step 1** to the tube containing charged resin. Incubate at room temperature for 5 min with gentle shaking. Spin at 1,000 × *g* for 1 min. Remove the supernatant with a pipette.
5. Repeat **step 4** until all supernatant has been incubated with the resin.
6. Wash the resin with 300 μL of 1× binding buffer, three times.
7. Wash the resin with 300 μL of 1× wash buffer (part of the His-Bind kit), twice.
8. Elute bound protein with 300 μL of 1× elute buffer (part of the His-Bind kit). Repeat the elution and combine two elutes for a total volume of 600 μL. The final elute can be temporarily stored at 4 °C for dialysis and quantification.

3.3 Preparation of RGTA-RGTB1 Rab-GGT Complex

This subhead describes the procedure of purifying up to 50 μg of heterodimeric Rab-GGT from 250 mL yeast culture. The α-subunit and β-subunit are co-expressed and form a heterodimeric complex in the same cell, and each subunit is tagged with a unique epitope tag. The complex is pulled down by a monoclonal anti-FLAG antibody coupled with agarose beads, and is eluted by competitive binding of 3× FLAG peptide [19]. Yeast cells can be difficult to break, due to the presence of a rigid cell wall. This subheading also describes the yeast protein extraction procedure by removing cell wall with lyticase digestion, followed by mechanical disruption of the spheroplasts [20].

1. Inoculate a single colony of YPH499 pESC-HIS RGTB1-FLAG RGTA1-cMyc yeast strain to 5 mL of SD-His medium. Shake overnight at 30 °C.
2. Inoculate all 5 mL of overnight culture into 50 mL of SD-His medium. Shake at 30 °C for 2 days (48 h).

3. Pellet the cells by centrifugation at 5,000 × *g* for 5 min, and discard the supernatant. Wash the pellets by resuspension with 20 mL sterile water, followed by centrifugation at 5,000 × *g* for 5 min, and discarding the supernatant. Repeat the wash twice for a total of three washes to completely deplete glucose (*see* **Note 5**).
4. Resuspend the pellet with 5 mL SG-His medium and transfer the suspension to 250 mL SG-His. Shake at 30 °C for 16–24 h to induce the expression.
5. Harvest the cell by centrifugation at 5,000 × *g* for 5 min. Discard the supernatant and wash the pellet with 25 mL sterile water. Record the weight of the empty centrifugal tube/bottle and the weight of tube/bottle with washed pellet to determine the amount of harvested cells. Typically a culture described above may yield 2.0–2.4 g of cell pellet. The quantities given in the following steps are based on a 2.0 g pellet, please calculate accordingly for different amounts.
6. Freshly add 90 μL of 1 M DTT to 2.91 mL of the reaction buffer (part of the CelLytic Y Plus kit), to a final concentration of 30 mM. Resuspend the pellet in this solution. Save a 10 μL aliquot for monitoring the spheroplast formation.
7. Add 50 μL of 25 U/μL lyticase provided by the CelLytic Y Plus kit. Mix gently and aliquot the reaction mixture to microcentrifuge tubes.
8. Incubate at 37 °C for 30 min (*see* **Note 6**). Remove another 10 μL aliquot, measure the OD_{800} of the samples before (**step 6**) and after (**step 8**) digestion (*see* **Note 7**). Continue the incubation and check the OD_{800} of post-digestion mixture every 5–10 min, until it is 10–20 % of the pre-digestion reading.
9. Harvest the spheroplast by centrifugation at 1,500 × *g* for 5 min.
10. Add to 10 μL of 1 M DTT, 20 μL of 0.5 M EDTA pH 8.0, 100 μL protease inhibitor cocktail (Sigma-Aldrich, P8215; *see* **Note 8**) to 5 mL of extraction buffer (part of the CelLytic Y Plus kit). Resuspend the spheroplasts in this solution. Aliquot the suspension in 500 μL–1.5 mL microcentrifuge tubes containing 250 μL of glass beads.
11. Briefly vortex the tubes and disrupt the spheroplasts by vigorously shaking the tubes with a bead beater in cold room. Perform three 10-min disruptions with 5-min intervals in between.
12. Incubate the cell lysate at room temperature with gentle shaking for 20 min.
13. Pellet the insoluble debris by centrifugation at 12,000 × *g*, 4 °C for 15 min. Transfer the supernatant to clean tubes.

14. Further clarify the lysate by centrifugation at 8,200 × *g*, 4 °C for 10 min. Set the tubes on ice.
15. Gently mix EZView Red anti-FLAG beads to uniform slurry. For every 1 mL of supernatant from **step 14**, transfer 40 μL of the 50 % slurry into a 1.5 mL microcentrifuge tube on ice (*see* **Note 9**).
16. Add 500 μL of 50 mM TBS pH 7.4 to each tube containing anti-FLAG beads. Briefly vortex and spin at 8,200 × *g* for 30 s. Carefully remove the supernatant with a pipette. Repeat wash once as above, and set the tube on ice (*see* **Note 10**).
17. Add 1 mL of the supernatant from **step 14** to each tube containing washed beads. Incubate at 4 °C with gentle shaking for 2 h or overnight to maximize binding. After incubation, spin at 8,200 × *g* for 30 s, and remove the supernatant with a pipette.
18. Add 500 μL of 50 mM TBS pH 7.4 to each tube with beads. Vortex briefly and incubate with gentle shaking at 4 °C for 5 min. Spin down the beads at 8,200 × *g* for 30 s, and remove the supernatant with a pipette. Repeat the wash as above twice for a total of three washes.
19. For every 40 μL slurry aliquot in **step 15**, freshly prepare 3× FLAG elution buffer by adding 3 μL of 5 μg/μL 3× FLAG peptide stock solution to 100 μL of 50 mM TBS pH 7.4.
20. Add 100 μL of 3× FLAG elution buffer to each tube with beads. Incubate with gentle shaking at 4 °C for 30 min. After incubation, spin down the beads at 8,200 × *g* for 30 s. Combine all of the supernatants and transfer to a clean tube. The final elute can be temporarily stored at 4 °C for dialysis and quantification.

3.4 Diafiltration, Concentration and Quantification of the Purified Proteins

The immediate elutes from the purification process contain excessive elution agents that may interfere with quantification and/or enzymatic reaction, such as glutathione, imidazole, and FLAG peptide. Amicon Ultra centrifugal filter provides a fast and easy way to exchange the original elution buffer with a neutral buffer environment without any interfering impurities [21]. The filter also helps concentrate the relatively low-yield proteins.

1. Check the induction and extraction efficiency, as well as the size of desired proteins by SDS-PAGE and/or Western blot.
2. Insert an Amicon Ultra-0.5 centrifugal filter into a collection tube provided along with the filters. Add 500 μL of purification elute to the filter and snap close the cap. Spin at 14,000 × *g*, 4 °C for 10 min. The remaining volume in the filter should be around 50 μL (*see* **Note 11**). Discard the flow through in collection tube. If the elute volume is more than 500 μL, add the rest to the filter and spin for another few minutes.

3. Add 450 μL of diafiltration buffer to the filter, and mix by pipetting up and down a few times. Insert the filter back into the collection tube, cap it, and spin at 14,000×*g*, 4 °C for 10 min. Discard the flow through. Repeat the diafiltration once.
4. Insert the filter upside down into a fresh collection tube (*see* **Note 12**). Spin at 1,000×*g* for 2 min. Transfer the concentrated solute to a clean 1.5 mL microcentrifuge tube. For proteins highly expressed in *E. coli*, add 250 mL of diafiltration buffer to the concentrate for dilution if necessary.
5. Quantify the concentration of each protein sample with NanoDrop or any Bradford assay-based colorimetric quantification. Adjust the concentration to that which is desired by adding diafiltration buffer or concentration with Amicon filters.
6. The purified and dialyzed protein can be stored at 4 °C for several weeks to months. For long-term storage, snap-freeze the tubes in liquid nitrogen and store in −80 °C freezer (*see* **Note 13**).

3.5 Prenylation Reaction and Detection

This subheading describes the procedure of conducting the prenylation reaction, separation of components by SDS-PAGE, and detection by autoradiography, for qualitative or semiquantitative analysis. The protein gel is treated with autographic intensifier due to the low-energy emission of the labeling isotope ^{3}H [22].

1. Assemble the reaction mixture (25 μL) in a 1.5 mL microcentrifuge tube as follows (*see* **Note 14**):

   5 μL of 5× Rab-GGT reaction buffer.

   1 μL of RGTA1-RGTB1 complex.

   2 μL of RabA4b.

   1 μL of AtREP.

   1 μL of [^{3}H]-GGPP.

   15 μL of sterilized water.

   Incubate at 30°C for 30 min.
2. Terminate the reaction by adding 5 μL of 6× protein sample loading buffer, and mix well. Incubate at 98 °C for 1 min.
3. Split the reaction mixture in half, and add each half to an SDS-PAGE gel. Run the two gels in the same electrophoresis cell until the dye reaches the bottom. Stain one gel with any Coomassie-based staining technique to visualize all the proteins. The other gel will proceed to autoradiographic detection as follows.
4. Transfer the gel to a 10×10 cm square plate. Add 5 mL gel fixation buffer and shake on a platform shaker for 30 min.

5. Rinse the fixed gel with continuous tap water for at least 15 min to completely remove acid residue (*see* **Note 15**).
6. Transfer the gel to another clean square plate. Add 20 mL Autofluor autoradiographic intensifier and 200 μL of 50 % (v/v) glycerol, and shake on a platform shaker for 30 min (*see* **Note 16**).
7. Do not rinse the gel. Instead, directly transfer the gel onto two layers of filter paper. Cover the gel with a layer of plastic wrap. Gently press out any air bubbles between the gel and filter paper. Dry on the gel dryer with heat and vacuum for 2 h. Continue vacuum without heat for another 30 min.
8. Remove the plastic wrap and the bottom layer of filter paper (*see* **Note 17**). In a darkroom, directly lay a piece of XAR film onto the gel surface, close and securely lock the cassette. Place the cassette in a −80 °C freezer for exposure.
9. Perform the first exposure for 24 h. Develop the film and adjust the exposure time according to the band intensity (*see* **Notes 18** and **19**).

4 Notes

1. The IPTG concentration and temperature for induction here are optimized for the constructs used in this protocol. If the protein yield is low for other constructs, try increasing the IPTG concentration and/or induction temperature (no more than 37 °C). If the desired protein tends to insoluble, lowering the IPTG concentration and/or induction temperature may help.
2. The expected result is to have a clear lysate. Start with lower energy output and shorter pulse, with enough intervals between pulses. Keep the tube on ice and avoid foam formation to prevent the desired protein from being heated and denatured.
3. Optionally, the N-terminal GST tag can be removed by thrombin cleavage. Although our results show that the GST tag does not interfere with prenylation reaction, removal may be preferred to show the original size of the desired protein, due to the considerable size of GST tag. A Thrombin CleanCleave kit (Sigma-Aldrich, RECOMT-1KT) works well for the immediate elute from SpinTrap kit purification.
4. Strong reducing agents such as DTT in the cell lysate reduce the Ni^{2+} ion, resulting in brown precipitation and low binding efficiency of His-tagged proteins. If the desired proteins require a reducing environment, use milder reducing agents such as β-mercaptoethanol (final concentration up to 20 mM), or those compatible with nickel-based His-tagged protein purification, such as TCEP-HCl.

5. The promoter of the pESC-HIS vector is inhibited by glucose, and the inhibition is relieved when the carbon source is switched to galactose. It is very important to make sure that the induction media is glucose-free for efficient protein expression. Wash the cell pellet with enough water and decant the supernatant thoroughly.
6. The time need for efficient lyticase digestion for our yeast strain is longer than that specified in the manufacturer's instructions of the CelLytic Y Pro kit. If using a different yeast strain, start monitoring the degree of spheroplast formation earlier and check more frequently, to determine the time needed. Longer digestion is discouraged as it may result in low yield and/or more denatured protein.
7. Make proper dilution of digestion samples to ensure that the readings fall in the effective and accurate measurement range of the spectrophotometer used. When comparing the readings, make sure that they are of the same dilution ratio.
8. The solvent of protease inhibitor cocktail is DMSO, with a higher melting point (19 °C) than water. Aliquot the cocktail right after it completely thaws. Do not set thawed cocktail on ice before aliquot, otherwise it refreezes and needs to thaw again.
9. To smoothly aliquot and disperse the beads, cut 1–2 mm off the tip of regular pipette tip using a clean razor blade or scissors.
10. The beads can be pooled together for the wash before aliquoted for protein binding. The volume of 50 mM TBS needed for each wash should be no less than 20 times of total packed gel volume. Note that the bead content in the original slurry is 50 %; thus the packed gel volume is only half of slurry volume [19].
11. Once the membrane in Amicon filter is wet, do not leave it dried out. Add diafiltration buffer immediately after the previous spin is done.
12. For maximum recovery, do not leave the concentrate in the filter for long. Do the reverse spin immediately after diafiltration [21].
13. Freeze–thaw cycle significantly affects protein activity in solution. Aliquot the dialyzed protein in small volume to multiple tubes before snap-freeze, only thaw one of the tubes at a time.
14. Mix all components except [^{3}H]-GGPP first to reduce radioactive waste generation. Add [^{3}H]-GGPP last and gently tap the tube to mix.
15. The acid residue in the gel must be completely rinsed out before Autofluor treatment; otherwise it will spoil some of the chemicals in Autofluor. Stopping the electrophoresis in

Subheading 3.5, **step 3** before the dye runs out of gel is a good idea—the remaining dye in the gel is a good pH indicator, which turns yellow during fixation (acid treatment) and turns back blue when the gel is rinsed to neutral.

16. Adjust the time of Autofluor treatment accordingly (30 min per 1 mm of gel thickness) if a thinner/thicker gel is used.
17. After vacuum drying, the gel will be in a light tan color similar to that of fresh Autofluor reagent. The appearance is sparkling like a layer of fresh fallen snowflakes [22].
18. The gel and cassette will draw moisture after being taken out of the freezer. Always leave the cassette open overnight in a dry environment to allow the gel to completely dry again before inserting film for another exposure of the same gel.
19. The addition of glycerol in Subheading 3.5, **step 6** should help prevent the gel from cracking during exposure under extreme low temperature. If the gel still cracks and leaves crack marks on the film, another exposure with a fresh film should make a clean autoradiograph.

References

1. Zhang FL, Casey PJ (1996) Protein prenylation: molecular mechanisms and functional consequences. Annu Rev Biochem 65:241–269
2. Randall SK, Marshall MS, Crowell DN (1993) Protein isoprenylation in suspension-cultured tobacco cells. Plant Cell 5:433–442
3. Yang Z, Cramer CL, Watson JC (1993) Protein farnesyltransferase in plants. Molecular cloning and expression of a homolog of the β-subunit from the garden pea. Plant Physiol 101:667–674
4. Yalovsky S, Loraine AE, Gruissem W (1996) Specific prenylation of tomato Rab proteins by geranylgeranyl type-II transferase requires a conserved cysteine-cysteine motif. Plant Physiol 110:1439–1359
5. Crowell DN (2000) Functional implications of protein isoprenylation in plants. Prog Lipid Res 39:393–408
6. Maurer-Stroh S, Eisenhaber F (2005) Refinement and prediction of protein prenylation motifs. Genome Biol 6:r55
7. Zeng Q, Running MP (2008) Protein lipid modification and plant development. Floriculture Ornamental Plant Biotechnol 5:319–328
8. Running MP, Lavy M, Sternberg H, Galichet A, Gruissem W, Hake S, Ori N, Yalovsky S (2004) Enlarged meristems and delayed growth in *plp* mutants result from lack of CaaX prenyltransferase. Proc Natl Acad Sci USA 101:7815–7820
9. Cutler S, Ghassemian M, Bonetta D, Cooney S, McCourt P (1996) A protein farnesyl transferase involved in abscisic acid signal transduction in *Arabidopsis*. Science 273:1239–1241
10. Running MP, Fletcher JC, Meyerowitz EM (1998) The *WIGGUM* gene is required for proper regulation of floral meristem size in *Arabidopsis*. Development 125:2545–2553
11. Johnson CD, Chary SN, Chernoff EA, Zeng Q, Running MP, Crowell DN (2005) Protein geranylgeranyltransferase I is involved in specific aspects of abscisic acid and auxin signaling in *Arabidopsis*. Plant Physiol 139:722–733
12. Hála M, Soukupová H, Synek L, Zárský V (2010) *Arabidopsis* RAB geranylgeranyl transferase β-subunit mutant is constitutively photomorphogenic, and has shoot growth and gravitropic defects. Plant J 62:615–627
13. Zeng Q, Wang X, Running MP (2007) Dual lipid modification of *Arabidopsis thaliana* Gγ-subunits is required for efficient plasma membrane targeting. Plant Physiol 143: 1119–1131
14. Hála M, Eliás M, Zárský V (2005) A specific feature of the angiosperm Rab escort protein (REP) and evolution of the REP/GDI superfamily. J Mol Biol 348:1299–1313
15. Leung KF, Baron R, Seabra MC (2006) Thematic review series: lipid posttranslational modifications. Geranylgeranylation of Rab GTPases. J Lipid Res 47:467–475

16. Caldelari D, Sternberg H, Rodríguez-Concepción M, Gruissem W, Yalovsky S (2001) Efficient prenylation by a plant geranylgeranyltransferase-I requires a functional CaaL box motif and a proximal polybasic domain. Plant Physiol 126:1416–1429
17. GE Healthcare (2009) Product booklet, GST SpinTrap columns
18. EMD Biosciences, Inc. (2006) User protocol TB054 Rev. F 0106, Novagen His Band Kits
19. Sigma-Aldrich, Inc. (2008) Technical bulletin, EZview Red ANTI-FLAG M2 affinity gel
20. Sigma-Aldrich, Inc. (2011) Technical bulletin, CelLytic Y plus kit for enzymatic lysis of yeast
21. Millipore Corporation (2011) User guide, Amicon Ultra-0.5 centrifugal filter devices
22. National Diagnostics (2008) Product protocol, Autofluor autoradiographic image intensifier

INDEX

Mark P. Running (ed.), *G Protein-Coupled Receptor Signaling in Plants: Methods and Protocols*, Methods in Molecular Biology, vol. 1043, DOI 10.1007/978-1-62703-532-3,

© Springer Science+Business Media, LLC 2013

MIX
Papier aus verantwortungsvollen Quellen
Paper from responsible sources
FSC® C105338

If you have any concerns about our products,
you can contact us on
ProductSafety@springernature.com

In case Publisher is established outside the EU,
the EU authorized representative is:
Springer Nature Customer Service Center GmbH
Europaplatz 3, 69115 Heidelberg, Germany

Printed by Libri Plureos GmbH
in Hamburg, Germany